The Second Severn Crossing

Ashford International Passenger Station

LAING INVESTORS
CONSTRUCTORS OF INFRASTRUCTURE

With 30 years experience of infrastructure development and investment worldwide, Laing will continue to build its portfolio of Infrastructure projects.

Our experience includes:
**Ashford International Passenger Station
The Second Severn Crossing
Midland Metro Line 1
Manchester Metrolink
Chiltern Railways
The Spanish Europistas and
Eurovias Toll Roads**

Manchester Metro Link

John Laing Construction Ltd
Amber House
Wood Lane
Hemel Hempstead
Hertfordshire
HP2 4TP

Contact: Katherine Falconer

Tel: 01442 65566
Fax: 01442 219844

Engineered for Success

Contact:
Roger Buckby, Bridges Director
Halcrow
Burderop Park, Swindon
Wiltshire SN4 0QD

Tel: 01793 812479
Fax: 01793 812089

HALCROW

Working Together To Achieve Environmental Solutions

Consultants to Laing-GTM and Welsh Office for the Second Severn Crossing

Providing:

- Environmental Impact Assessment
- On-site Environmental Liaison
- Dust and Noise Monitoring
- Ecological Monitoring
- Checking Pollution Control Measures
- Advice on Restoration
- Environmental Management Systems

For further information please contact Sue Lees at SGS Environment, Ivor House, Bridge Street, Cardiff CF1 2EE
Tel: 01222 664684, Fax: 01222 664768.
Please quote reference CES/08/97

SGS Environment

HYDRATIGHT
The Jacking Specialists

MORLIFT offer the total solution in the specialist area of lifting, lowering, moving and weighing heavy and complex structures whether it is sales, service or hire.

Hydra-Tight Limited - **MORLIFT** BENTLEY ROAD SOUTH, DARLASTON, WEST MIDLANDS, WS10 8LQ

0121 50 50 600
FAX: 0121 50 50 800

MORLIFT

PROCEEDINGS OF THE INSTITUTION OF CIVIL ENGINEERS

Second Severn Crossing

SUPPLEMENT TO CIVIL ENGINEERING
VOL. 120 • SPECIAL ISSUE 2 • 1997

CONTENTS

Introduction. *B. P. Pritchard.*	2
Second Severn crossing—initial studies. *P. R. Head, P. Iley and E. T. Bradley.* Paper 11439	3
Second Severn crossing—pre-construction period and design development. *J. N. Kitchener and D. H. Mizon.* Paper 11440	13
Second Severn crossing—design and construction of the foundations. *J. N. Kitchener and S. J. Ellison.* Paper 11441	22
Second Severn crossing—viaduct superstructure and piers. *D. H. Mizon and J. N. Kitchener.* Paper 11442	35
Second Severn crossing—cable-stayed bridge. *D. H. Mizon, N. Smith and A. J. Yeoward.* Paper 11443	49
Second Severn crossing—management of construction. *J. N. Kitchener and D. H. Mizon.* Paper 11438	64

Front cover:
Second Severn Crossing

Introduction

B. P. Pritchard, BSc, MS, FICE, FIHT

Welcome to this special issue of *Civil Engineering*, the general journal of *Proceedings of the Institution of Civil Engineers*. It is intended to provide an overview of the planning, design and construction of the award-winning second Severn motorway bridge, one of the largest design, build, finance and operate projects undertaken in the UK and another triumph of Anglo-French collaboration.

This is the second special issue of 1997 and the first to be dedicated to a bridge project. Previous special issues on crossings—four on the English Channel and one on Storebælt—were about tunnelling. Indeed, various tunnelling options were examined for the second Severn crossing, but the elegant cable-stayed bridge and viaduct design ultimately proved to be the optimum solution.

The first paper in this special issue is by Peter Head, Peter Iley and Ed Bradley. It describes the planning and background to the £330 million construction project and how it eventually formed part of a £1 billion concession for the operation and maintenance of both Severn motorway crossings for 30 years.

Neil Kitchener and David Mizon then discuss how the successful DBFO bid was put together by the Anglo-French partnership of Laing and GTM, and report on how design ideas and construction methods were developed to produce the final design concept.

The next paper by Neil Kitchener and Stephen Ellison covers the detailed design and construction of the foundations for the 49 viaduct supports and two cable-stayed bridge pylons. Due to strong currents and a high tidal range, 33 of the foundations were built using precast caissons weighing up to 2000 t. Particular care had to be taken where the viaduct crossed an existing railway tunnel.

Construction of the two 2 km viaduct structures involved match casting over 1100 deck units on site and use of a purpose-built 234 m long erection gantry. This is described in the following paper by David Mizon and Neil Kitchener, who also cover M&E works including the monorail deck access train.

David Mizon, Neil Smith and Andrew Yeoward then report on the detailed design and construction of the 456 m span cable-stayed bridge at the centre of the crossing. The 900 m long composite steel and concrete deck is supported from two 194 m high pylons with 240 cables. The paper also explains how excessive cable oscillations were successfully dealt with prior to opening.

Finally, Neil Kitchener and David Mizon describe how construction of the project was managed. With equal managerial input from the British and French partners, the project was completed on time, within budget and without loss of life or limb.

My only regret is that the papers do not record my early struggle, on behalf of my fellow Welshmen, to locate the toll booths on the Welsh side!

My thanks go to all the authors for their time and patience in preparing and revising their papers, to the referees for reviewing them and to Neil Kitchener at John Laing in particular for coordinating the production and delivery of all papers on the ICE's behalf.

Please note that all papers will be presented and discussed at an ordinary meeting at the Institution on 2 December 1997 at 2p.m., which is free for all to attend. Written discussion contributions will need to be submitted no later than one week after the meeting.

Brian Pritchard, consultant and formerly bridge director of W S Atkins, is a member of the ICE Civil Engineering Panel and was responsible for the assessment of all papers in this special issue.

Second Severn crossing— initial studies

P. R. Head, FEng, FICE, FIStructE, FIHT, P. Iley, CEng, FICE, FIHT, and E. T. Bradley, BSc, MS, CEng, MICE, MIHT

Construction of the £330 million second motorway bridge across the Severn Estuary between England and Wales is part of a package of over £1 billion which includes new motorway approach roads and the financing, operation and maintenance of both crossings. This paper describes the background to the second crossing—from the initial feasibility study started in 1984 and subsequent geotechnical, hydraulic and aerodynamic investigations to the preparation of tender documentation, appointment of the concessionaire in 1990 and start of the 30 year concession following parliamentary and planning approval in 1992.

In February 1984 the UK Government announced that a study was to be commissioned into how a second motorway crossing of the Severn estuary might be provided in the general corridor of the existing M4 crossing. The Government's Department of Transport (DoT) invited tenders from engineering consultancies to undertake these studies and the successful bid was from the Second Severn Crossing Group (SSCG), comprising consultants G. Maunsell & Partners and W. S. Atkins. The SSCG was appointed in August 1984.

The consultants' brief required that the study should consider three forms of fixed crossing

- bridges
- driven (bored) tunnels
- immersed tunnels.

The limits of the study area (Fig. 1) were suggested to be approximately 8 km upstream and downstream of the existing Severn road bridge although crossing locations outside these limits were not precluded.

Traffic

The annual average daily traffic (AADT) crossing the bridge in 1984 was just over 35 000 vehicles (two-way). By 1996, the assumed earliest date for opening, this was estimated to increase by 32% on a high-growth assumption and 14% on a low-growth assumption. Taking 2010 as the design year, these figures project to 56% and 25% respectively, giving estimated traffic of around 55 000 and 44 000 AADT. Summer weekend holiday traffic has been measured at up to 48% above these levels.

In practice traffic grew strongly—by almost 30%—with economic growth between 1986 and 1990, increasing pressure for an additional crossing.

Although the existing crossing is classified as motorway, there are not full hard shoulders on the bridge. The crossing's design capacity can therefore be considered to be closer to that of a dual two-lane all-purpose road, at up to 46 000 AADT, rather than a motorway.

The crossing's capacity is further restricted for about 130 h per year during periods of high winds owing to its exposed nature. These restrictions take the form of reduced speed limits or restrictions on high-sided vehicles but can, in extreme weather conditions, result in total closure of the crossing.

Approximately 75% of the traffic crossing the existing bridge was known to have an origin or destination in the Newport/Cardiff/Swansea direction, with only 25% having a desire to go to the north for the Chepstow/Wye Valley area.

An early decision to provide a second crossing with, if possible, substantial windshielding and to a higher standard of provision made it desirable that the majority of traffic would use the new crossing, preferably by choice rather than traffic management. Fig. 1 shows that any location for the new crossing downstream of the existing bridge would also bring a reduction in journey distance for the 75% of traffic heading to or leaving South Wales. It was therefore possible from an early stage of the study to concentrate on locations downstream of the existing bridge.

A second crossing, and especially a bridge, located immediately adjacent to the first would have minimized the lengths of new roads to connect it to the existing road system. Factors against this however, were

Peter Head is project director for the Government agent

Peter Iley is project manager for the Government agent

Ed Bradley was project manager for the Department of Transport

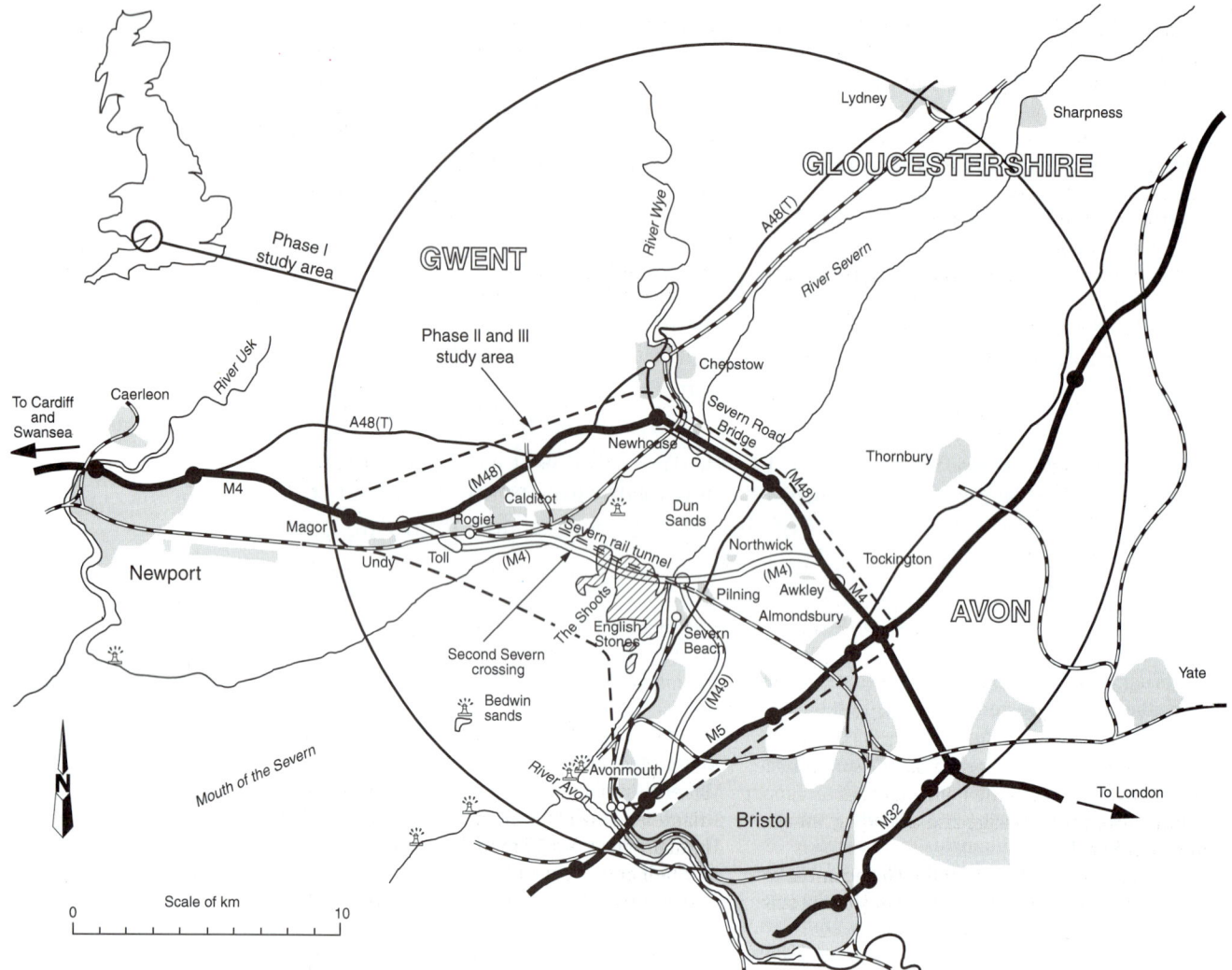

Fig. 1. Study area

- development which had taken place immediately adjacent to the existing bridge since its construction
- the necessity for aesthetic and navigation reasons to make the second bridge as near as possible a duplication of the first, which would not necessarily be the most economical solution
- the possible aerodynamic effects on the existing bridge caused by an adjacent structure and vice versa
- the risk of interaction between the foundations for the new and existing bridges
- the first bridge, not surprisingly, already occupied the best location.

Figure 1 also shows that development on the Welsh shore inhibits the choice of any routes for about 5 km downstream and that this location coincides with an extensive area of bedrock exposed at low tide for almost half the width of the estuary, known as the English Stones.

Although experience indicated that tunnels are invariably more expensive than bridges they do have three clear advantages and had to be considered. A tunnel is visually less obtrusive than a bridge. A tunnel also has the advantage that one tunnel can be constructed initially to take one carriageway with a second tunnel added alongside it at a later date when traffic demands require it—a feature more difficult with a bridge (phased implementation of the second crossing was part of the study brief). Furthermore, tunnels are less susceptible to adverse weather conditions, a significant factor in this location.

From a strictly engineering aspect both bored and immersed tunnels were feasible, although not in the same location as they have different requirements. A bored tunnel was not only feasible but proven, as the chosen route for the second crossing actually goes over the existing rail tunnel taking the main London/Cardiff railway under the estuary. The exposed surface bedrock at that site would not favour an immersed tunnel solution. However, further upstream, between the English Stones and the existing bridge and also further downstream, estuary bed conditions are more favourable for an immersed tunnel.

The main factor against either tunnel solution compared with a bridge was cost, both capital and operational (ventilation, drainage, lighting), even for the dual two-lane carriageways under consideration at that time. The subsequent decision to construct the crossing to dual three-lane standard with hard shoulders would have made the cost prohibitive for a bored tunnel, not only for the area of unusable space in a large circular bore but also the depth to which a large tunnel would need to be placed to provide adequate cover.

With a bridge favoured for the English Stones site, the connections to the existing M4 on the Welsh side had few major options although refinement to this line continued right up to tender in 1989.

On the English side, consideration was primarily given to connecting to the M4 west of the M4/M5 interchanges at Almondsbury. As the study developed, an additional connection to the M5 at Avonmouth was found to offer substantial savings in distance and be economically worthwhile.

As both the M4 and M5 in this area are already reaching saturation at holiday peak times the final decision was to provide connections to both roads, with the additional benefit of prolonging the life of the lengths of the existing motorways which were duplicated.

This then was the scheme finally recommended to the Government in the consultant's final report in July 1986, with a bridge and a road layout as shown in Fig. 1.

The decision was taken by the Government to proceed with this scheme and to consider procuring the project itself by a design and build (DB) contract. In 1987 tenders were invited from consultants to develop the scheme and the SSCG was again appointed. Also in 1987, the Government's management of the project moved from the London bridge engineering division (BE) to the south-west regional office (SWRO) at Bristol. A project steering group was set up, chaired by the SWRO director, and included representatives from W. S. Atkins, G. Maunsell, the Welsh Office and the DoT (Tolled Roads and Crossings (TRC) and BE). One of the first tasks of the SWRO was to organize a series of public exhibitions at local venues in England and Wales to explain the proposed scheme and obtain the views of the public; this was to have a significant effect on the successful development of the scheme allowing concerns of the public, together with industry and organizations such as British Rail and Gloucester Harbour Trustees and their pilots, to be taken into account.

Between 1987 and 1989, and before any fixed decision had been made on the form of contract, the consultant's design for the main river crossing was developed sufficiently to establish the main design parameters, specification requirements and a robust illustrative design. This meant that the tender period could be set at six months because the preliminary investigations had increased confidence in the route and the outline design, thereby also minimizing tenderers' costs.

Between the beginning of 1987 and the issue of the tender documents in April 1989 significant reports were prepared, half of these by the SSCG themselves and the remainder by specialist organizations commissioned by the SSCG on behalf of the Government. The majority of these reports fall into one of three main groups

- ground investigations
- hydraulic investigations
- aerodynamic investigations.

Ground investigations

Ground investigations were carried out between July 1988 and February 1989. At about this time, consideration was being given to the procurement of the main crossing by design and build tenders and to proceeding with the approach roads by conventional consultant design and site supervision using the standard ICE 5th edition contract.

The aim of this early survey work for the bridge was to provide data at the main pier positions and give guidance on the bridge viaduct approaches, but not in fine detail since the contractors' viaduct pier positions could and did vary.

For the main crossing, a grid of cores was taken across the estuary. These were 100 mm in diameter down to around 15 m depth and 75 mm in diameter from 15 m to 40 m depth.

Shell and auger holes appropriate to this stage were sunk along the line of the approach roads, supplemented by a more extensive electric cone penetrometer survey. The auger holes were extended by rotary coring into bedrock where necessary or relevant at bridge sites. These investigations were carried out by Soil Mechanics Ltd.

The data was supplemented by a geophysical survey of the estuary carried out in late 1988 by Electronic & Geophysical Services Ltd. Both sonar buoy and bottom drag cable techniques were tested during trials and over 40 km of marine refraction traversing was eventually completed using the latter system. These produced sea-bed profiles and contours together with geological data which differentiated between layers of sedimentation and various types of bedrock along a series of sections.

Additional marine refraction traversing was undertaken in the vicinity of the expected crossover point above the Severn rail tunnel to increase data for pier foundation design.

A more detailed site investigation to be carried out later by the successful contractor was written into the Government's requirements for the bridge.

Hydraulic investigations

The first of the hydraulic studies was carried out in 1985 as part of the main feasibility study and consisted of a numerical model to assess the

Fig. 2. 1:225 natural-scale model of part of the estuary showing line of caissons either side of the main channel

effects of various crossing options on the tidal regime of the estuary. A numerical model was chosen as being a cheap, quick and flexible means of testing a large number of bridge and tunnel options over some 6 km length of estuary.

Following selection of the preferred crossing site at the English Stones, it was necessary to establish in more detail the impact of the new bridge on the main navigation channel, known as The Shoots, which passes through the main span of the bridge.

Following Hydraulics Research Ltd's previous work with the numerical model, that company was engaged for the additional studies and recommended a physical model of a more localized area covering a 2·5 km length of estuary mainly upstream of the bridge and from the Welsh shore to approximately two-thirds of the width of the estuary. This area was decided on after discussions with the navigation authority, Gloucester Harbour Trustees, which was particularly concerned about outward-bound shipping movements and the difficulty in manoeuvring into the relatively narrow navigation channel amid strong tidal currents.

The large variations in bed level, including the 30 m deep gorge of The Shoots, caused problems with the accuracy of the numerical model, as current speed and direction at bed level are frequently very different from at the surface. It was therefore decided that the physical model could be to a natural scale, and 1:225 was chosen (Fig. 2).

Although the estuary has the second highest tidal range in the world, at 14·5 m, outward-bound shipping only passes the location of the bridge between around high water + 1 h and high water + 2 h. The model was therefore run on a non-tidal basis at these two ebb tide states.

Having established the effect of the crossing on the tidal flows through The Shoots it was considered necessary to undertake a more detailed examination of ship handling in the channel. The use of radio-controlled vessels on the physical model was discounted on the basis of the difficulty in modelling manoeuvring in non-real time. The decision was therefore taken to carry out a ship simulator study, utilizing data derived from both the numerical and the physical models. This work was carried out by Maritime Dynamics Ltd and included 'hands on' experience with the simulator by pilots of Gloucester Harbour Trustees. The brief was extended to examine a range of failure situations so that the risk to drifting outward-bound vessels could be evaluated.

The ensuing risk assessment model allowed the determination of the probability of ships of vari-

INITIAL STUDIES

ous sizes impacting on piers at various locations in the estuary.

The outcome was that there was seen to be a high probability that an errant outward-bound vessel would drift to the west of the main span by up to 800 m with a possibility of striking one of the eight piers in that length of bridge. The piers were designed to withstand impact from the largest vessels which can use that stretch of river but the harbour authority was also concerned at the impact effect on the vessels themselves. With a pier width/span ratio of 1:10 and a high current, the effective reduction in flow width was already greater than 10% so additional protection around the piers was not desirable. The harbour authority proposed low rubble islands upstream and in line with each of the eight piers and these were incorporated in the notional design, but were never built, as will be discussed in a later paper. A secondary outcome was the decision to provide, by a separate advance contract, four beacons to mark clearly the navigation channel through the bridge main span.

Aerodynamic studies

The third and final significant group of studies related to aerodynamics. The consultants' original brief required analysis of all crossing options with respect to the effect of high winds on traffic (or on the unavailability of the crossing due to problems with traffic subject to high cross-winds). The protected environment of tunnel solutions was not viable on cost comparisons with bridges. It was therefore decided at an early stage of the initial study that the reliability of any bridge crossing would be improved as far as practicable by the addition of windshielding.

It was well known that the aerodynamic stability of bridge decks can be very sensitive to small changes in cross-section and, in particular, to edge detail. The addition of substantial edge parapets or windshielding was therefore likely to be of considerable significance aerodynamically.

Although it is feasible to protect fully the traffic on a bridge, a balance must be drawn between the cost of full protection, with the inherent problem of dynamic stability of a long structure, and providing sufficient protection to minimize disruption to traffic. In the event of closure of both bridges the alternative route by road is over 80 km longer.

The basic criterion adopted was to provide windshielding on the bridge such that traffic on the bridge would be subject to wind forces no

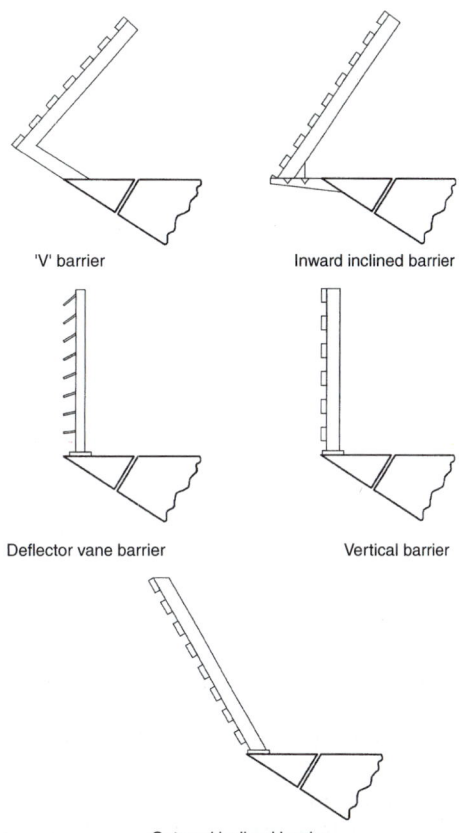

Fig. 3. Windshield configurations

Fig. 4. (a) Prototype windshielding; (b) computer image of windshielding

7

greater than those on the unprotected approach embankments; that is, if traffic could reach the bridge it could cross it.

The first series of tests, carried out by British Maritime Technology Ltd during the initial feasibility study, considered a series of vertical and inclined parapets as shown in Fig. 3, all of around 50% porosity and 3 m high. Because wind speed over the main part of the bridge will be approximately 1·4 times that on the approaches (because of the additional exposure) and the force is proportional to the square of the wind speed, the force on vehicles on a bridge without windshielding would be twice that on the approaches. Consequently, a reduction in wind speed, and hence porosity, of 50% is required. The first investigations, using 1:100 scale aeroelastic models, included 1000 m and 800 m span suspension bridges as well as a 450 m cable-stay bridge for evaluation of bridge types and studied, as well as barrier effectiveness, divergent amplitude responses, vortex shedding characteristics and drag coefficients for the various structures. Prototype wind speeds up to 60 m/s were specified.

These first investigations indicated that the aerodynamic response of the proposed structures was not affected significantly by the actual shape of windshielding.

A second series of tests, carried out by Atkins Research & Development, was therefore embarked on, using only the cheapest form of barrier (vertical) but experimenting with barrier heights of 2 m, 3 m and 4 m and also porosities of 45% and 40%, all to a larger scale of 1:50.

Following the derivation of a preliminary design configuration for the barriers from the above tests, the Motor Industry Research Association was engaged to carry out a series of full-scale tests on a prototype length of barrier (Fig. 4(a)) using their cross-wind generator and an instrumented high-sided van. The van was driven across the generated airflow before and after erection of the barrier and the benefit established. As a separate exercise the same vehicle was driven on roads local to the project under both calm and windy conditions.

Having proved the barrier configuration, a final series of studies and tests was carried out by Polar Oceans Associates to assess the performance of the barriers under ice and snow conditions likely to be encountered in the Severn Estuary. These indicated that the width of a 500 mm horizontal windshield plank might increase by up to 75 mm because of ice accretion and this was incorporated in the design requirements.

Specification

Most of the investigations were carried out in parallel and the results incorporated in the comprehensive specification, which was based on the Department of Transport's *Specification for Highway Works*, August 1986.

The preferred crossing was, at this stage, identified as being located at the English Stones and consisting of a 456 m main-span cable-stayed bridge over the navigation channel, connected to both shores by viaducts over 2 km long. Reliability, durability and maintainability were to be the key elements of the specification, or Government Requirements.

Investigation indicated that viaduct spans were likely to be between 90 m and 100 m in order to achieve an efficient balance between the costs of substructure and superstructure. Composite steel and precast concrete options for the decks were both feasible although the precast option was preferred owing to the greater knowledge of this form of construction at the time and an assumed economic edge.

At this time (1987/88) Maunsell was engaged in a UK nationwide study of the condition of highway bridges for the Department of Transport. Data from this study, together with experience from the offshore construction industry, was used in drawing up the specification for a durable concrete in an exposed environment. There was also historic evidence of problems in post-tensioned concrete due to incomplete filling of ducts with grout leading to tendon corrosion. The fundamental decision was therefore taken by the consultants and the bridge engineering division of the DoT to specify external tendons for precast concrete construction with a non-cementitious blocking medium as protection within the ducts. The system and the blocking medium were specified to be such that either individual strands or complete tendons could be withdrawn for examination and, if necessary, replaced. It was also specified that any single tendon could be removed without the need to impose any load restriction on the bridge.

A study was undertaken to compare the performance of conventional cement-grouted 'built-in' tendons and the unbonded external system. To compensate for the slightly reduced performance of unbonded tendons it is necessary to limit (reduce) their unbonded length or provide about 10% additional capacity. The former approach, with a maximum individual tendon length of 0·4 times the length of the span, was specified as studies indicated that the alternative of increasing the number of prestressing tendons would be more expensive.

For the cable stays of the main bridge, a similar protection system was specified, of an outer high-density polyethylene (HDPE) sheath and an internal blocking medium but, in addition, the strands were required to be galvanized.

All of the foregoing studies aimed to provide a clear, reliable technical framework which would allow maximum freedom for tenderers to develop cost-effective schemes rapidly during a relatively short six-month tender period. This had to be achieved within parameters which ensured that the Secretary of State's responsibilities and the requirements of affected and interested parties were fully respected.

The responsibilities included safety of the public using the crossing, shipping navigating the estuary, trains traversing the tunnel under the bridge, durability, maintainability, disruption to traffic during construction and keeping environmental effects to an acceptable level. This was in addition to ensuring that the proposals for the entire project were practical, offered acceptable economic return and would lead to deliverability of the approach roads, as well as the Severn crossing, through statutory processes, in this case by a parliamentary bill.

Illustrative design

It was agreed at the outset of the contract procedures that, if these objectives were to be achieved, a 'reference' or 'illustrative' design should be developed and issued as part of the tender invitation (Figs 5 and 6). There were numerous complex interfaces between the various disciplines and it was considered best practice in this sensitive location to build on the earlier work and provide a comprehensive document which gave guidance on how these matters could best be provided for. Tenderers would be unlikely to appreciate fully and solve all of the problems satisfactorily within the time available, unless they had optimum use and availability of the knowledge and time already invested by the Government and its consultant, the SSCG.

The illustrative design was the cornerstone of this guidance, but was not mandatory and still allowed freedom to tenderers within the framework of the Government's firm requirements. The illustrative design was of great benefit in discussions with third parties during tender preparation and later in comparing tenderers' alternative proposals. It went as far as to show a comprehensive access system of maintenance gantries and an innovative under-deck access train with frequent stations along the 5 km of bridge permitting access to the gantries or directly into the decks, thus enabling full inspection of the entire under-deck surfaces without requiring access from the carriageways and consequent interference to traffic.

In the event, the winning tender fully met the firm requirements and made best use of the illustrative design as well as being the most competitive. The finished bridge, in appearance and quality, closely matched that being sought by the Government, its consultant and the Royal Fine

Fig. 5. Illustrative design—cable-stayed bridge

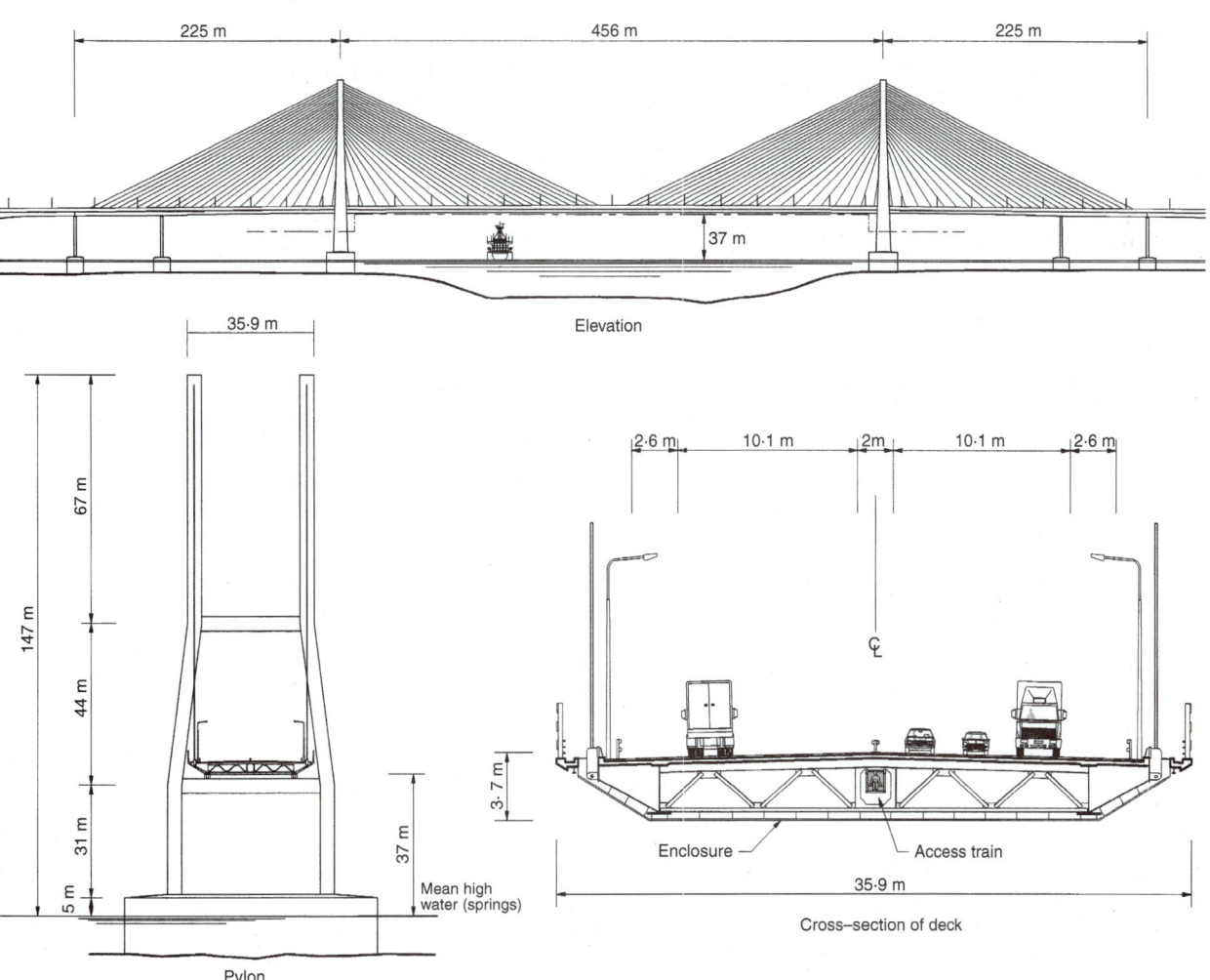

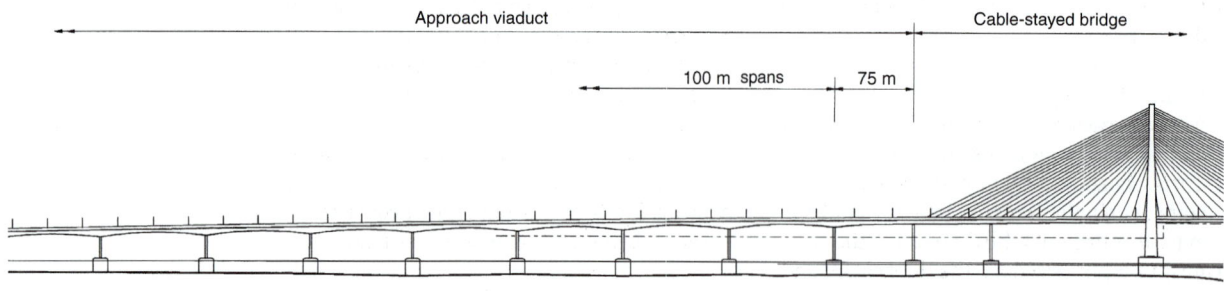

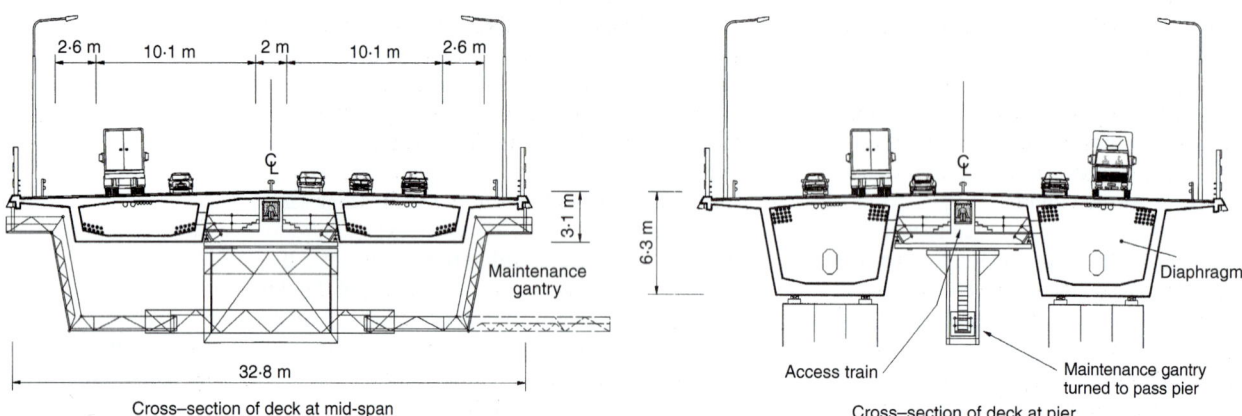

Fig. 6. Illustrative design—approach viaduct

Arts Commission, while also incorporating the design and build teams' ideas and innovation in construction techniques.

Tender invitation documents

The tender invitation documents comprised the following volumes
- Volume 1—administrative, contractual and financial aspects
- Volume 2—engineering and operational aspects
- Volume 3—studies and surveys
- Volume 4—legislation and other documentation
- Volume 5—illustrative design.

Tenderers were asked to tender for two options.

- *Option 1.* To design, build, finance and operate (DBFO) the second crossing, and to pay off the outstanding debts and maintain and operate the existing crossing, both under a limited-period concession arrangement with the Government.
- *Option 2.* To design and build the second crossing for staged payment by the Government.

The technical requirements were the same for both options and were based on the DoT *Specification for Highways and Bridges* but included major specific clauses required for this unique project, mainly derived from the previously mentioned studies, with the overall aim of achieving high quality, durability and maintainability.

An important innovation in the form of the technical requirements was the division of the documents into three series of back-to-back clauses

- firm requirements
- guidance
- requirements of the tenderers (in producing their tender).

Every specific, technical firm requirement clause was complemented by a separate clause under each of the other two headings.

The firm requirements, which also included those of third parties, were mandatory for a conforming tender. Guidance clauses gave the background to firm requirements.

Requirements of the tenderers set out the information which had to be submitted by them against each technical clause to demonstrate how each was being addressed. In assessing the tenders, these responses could be set side by side for comparison. Thus, non-conformances could be readily identified and referred back to the tenderers.

Tenders were invited in April 1989 with a six-month tender period followed by a further six-month period for tender assessment.

Tenders were compared on technical conformity by means of a spreadsheet database which was also used by the SSCG to report to the DoT. By this means the original four prequalified tendering consortia were reduced to two for further clarification and avoidance of ambiguities, if occurring.

Government agent

In 1990 the Government appointed G. Maunsell and Partners, under a separate commission, as its agent for the construction of the second crossing works embodied in the concession agreement. In doing so for this form of DBFO contract, the Government sought to reduce the duplication and extra cost often incurred with the conventional resident engineer approach, to place more reliance on quality assurance, and place more responsibility on the contractor, its designer and design checker as part of the technical approval procedures.

During the two-year period before the start of construction, the pre-concession period, the Government agent's role could be measured as follows, in many cases as representing the Government or as facilitator

- between the Government, concessionaire and third parties during the parliamentary processes
- between the contractor, third parties and landowners regarding commitments, access for surveys and further investigations
- disseminating acquired knowledge of the project to the concessionaire and contractor.

A further function was to monitor and approve payment by the concessionaire to the contractor or any other party of sums for design, testing, studies, procurement of plant and so on.

Only if so approved would the Government reimburse such sums in the event of failure of the bill or termination of the concession due to causes outside the concessionaire's control.

Concession agreement

The heads of agreement for the concession were made with Laing-GTM in April 1990 on the DBFO basis outlined. The concession agreement itself was signed in October 1990. The maximum term of the concession is 30 years but Laing-GTM and its financial backers Bank of America and Barclays de Zoete Wedd offered an alternative of a possible shorter term, depending on the payback period of the capitalized sum, and this was agreed. On acceptance, Laing-GTM and its backers set up an operating company, Severn River Crossing (SRC).

The signing of the heads of agreement in April 1990 triggered a number of operations which could proceed in parallel over the following two years.

- Preparation of the parliamentary bill could proceed. The input from Laing-GTM regarding the proposed construction methods, programme and environmental protection measures was invaluable at this stage in discussing the project with possible objectors to the scheme and with the select committees.
- Detailed design of the bridge could proceed, enabling vital decisions on methods of construction, capacity of lifting plant and so on to be made.
- Orders for the required plant and materials could be placed.
- Drawing together of the final concession agreement, as above, was possible.

The Severn Bridges Bill was put before Parliament in the 1990/91 session. The parliamentary bill was essential for re-establishing/establishing tolling on the existing and new crossings and it also short-cut the statutory process for new roads which normally requires draft orders and a public inquiry. On behalf of the Department of Transport, the SSCG and SRC produced a substantial, two-volume environmental statement during 1990. A full appraisal framework, as specified in the DoT's *Manual of Environmental Appraisals*, included full descriptions of the published scheme, the environmental and planning context, the contractor's construction proposals, their temporary effects and proposed mitigation measures and, finally, a description of the alternative schemes considered together with reasons for the choice of the published scheme. A two-volume landscape proposal report was also produced, with input from the Government's landscape advisory committee.

The study area contained a number of sites of special scientific interest (SSSIs), some of which were inevitably traversed by the published route. Hence, great care was necessary, not only in planning mitigation measures, but also in the control and monitoring of the construction phase. The Severn Estuary is now a special protection area (SPA) through the Ramsar agreement on sites of international importance and all major proposals and mitigation were agreed with the key environmental agencies.

The pre-bill environmental work carried out, including substantial landscaping proposals, was probably the most thorough investigation of this type ever carried out up to that time for a highways proposal.

After two readings in the House of Commons, where the principle of the proposal was accepted, a select committee sat for 10 days in March and April of 1991 to hear petitions against it. The principal petitioners and their main concerns were as follows.

- *Gloucester Harbour Trustees.* Navigation through the bridge and an additional anchorage.
- *British Rail.* Protection and safeguards to their tunnel where it was to be crossed by the bridge and the Avonmouth approach road.
- *Local parish councils.* Protection for local residents during construction and from traffic using the completed scheme.

- *Avon and Gwent county councils.* Level of tolling and improvements to the local roads for construction traffic.
- *Whitbread Brewery.* Safeguards for their water supply which comes indirectly from a freshwater aquifer and could have been endangered by boreholes, piling, etc.

As a result of the parliamentary process a number of undertakings were given to third parties and some minor amendments made to the published scheme.

The signing of the heads of agreement enabled the concessionaire to finalize a number of contracts. The start of the detailed design work enabled certain fundamental decisions to be taken regarding the maximum handling weight of major precast elements of the bridge and the corresponding plant required to move and place these elements. This in turn enabled the concessionaire to start procurement procedures for the necessary plant, much of which had to be purpose-built from components assembled from all quarters of the globe.

The fourth and last of the activities previously mentioned was the compilation of the final concession agreement. The main agreement comprised seven volumes, the first of which contained the main clauses of the contract between the Secretary of State for Transport and SRC. They were based on the principles outlined in the tender invitation and the concessionaire's proposals. After a considerable amount of work they were merged into one set of agreed clauses.

The remaining volumes encompassed twelve schedules. Typical of these was schedule 3 relating to the construction requirements for the second crossing. It had two parts. The first contained all the Government's requirements (or engineering specification) for the construction of the bridge as given in the tender invitation. The second part consisted of the concessionaire's proposals to comply with the requirements of part 1. Part 2 was nearly twice as large as part 1 and its clauses did not relate directly to those in part 1. This lack of direct reference and additional working made it more difficult to be certain of compliance during the tender assessment.

During the assessment every effort was made to ensure that the proposal met all the requirements, resulting in lengthy discussions and negotiations between parties between the spring and autumn of 1990. The requirements and the proposals were then incorporated into the agreement, which was signed by both parties. Whereas the vast majority of the agreement was clear, occasional inconsistencies were found subsequently. For the second bridge, these were all resolved on site, but for the existing bridge, the assistance of the disputes panel set up under the agreement was required to achieve a consensus over the extent of work covered in detail by the agreement.

The total financial package procured by SRC was valued at almost £1000 million to cover the three main aspects of bridge construction, payback of outstanding debt (over £120 million) on the existing bridge and financing costs.

In March 1997, having successfully completed the high-risk part of the concession deal, that is, having built the bridge, SRC was able to renegotiate its loan on significantly improved terms. This was hailed as a benchmark deal for the refinancing of private-finance-initiative infrastructure projects.

Pre-operation period

The period between the signing of the heads of agreement in April 1990 and the official start of the concession period following parliamentary approval was known as the pre-operation period. It had been expected that this period would be under two years but owing to slippage of a few weeks in Parliament and a general election in the spring of 1992 it was in fact about 25 months. This slight excess triggered certain compensation clauses.

The main significance of the pre-operation period was that any essential expenditure by the concessionaire or the contractors was underwritten by the Government so that, in the event of failure of the necessary legislation, compensation would be paid to SRC.

To control the expenditure SRC submitted monthly forecasts to the Government agent for scrutiny and verification before acceptance by the Government. By the end of the two-year period significant sums had been approved not only for major items such as marine plant but also for work actually carried out on site to improve local roads and construction accesses. This work on the local roads was largely made possible ahead of the Act because, particularly on the Welsh side, most of the land required for the improvements was owned by the county council.

The relative financial freedom resulting from the above arrangement ensured that no time was lost in starting construction after the legislation was passed. The bill received royal assent on 13 February 1992. The SSCG had prepared all the necessary land plans and schedules for the accompanying compulsory purchase orders so that by the first week in March the Government agent was able physically to take possession, on behalf of the Government, of all the land required for the construction of the crossing. The official start date for the concession period was 26 April 1992, which was also the first day of the four-year construction period.

Acknowledgements

The authors wish to acknowledge the contribution made by the following, particularly during the study phase of the project: the late Dr B. Richmond, Maunsell, project director 1984–1992; F. J. Parker, W. S. Atkins, project director 1984–1988; and P. Gosling, W. S. Atkins, project manager for the SSCG.

Second Severn crossing— pre-construction period and design development

J. N. Kitchener, BSc, CEng, FICE, *and D. H. Mizon*, CEng, MICE, MIStructE, MIHT, MIL

This paper describes the background to the award-winning Anglo-French design, build, finance and operate bid for the £330 million second Severn motorway bridge between England and Wales. It explains how the tender design was prepared, reviews the commercial and managerial implications of the 30 year concession agreement and reports on how design and construction methods were developed to produce the final design.

During 1989 UK contractor John Laing Construction and French construction company GTM Entrepose joined forces to bid for the contract for the second Severn motorway crossing between England and Wales. The two companies have a history of collaboration and joint venturing going back over 30 years, particularly on pipelines and oil platform construction.

It was always intended that the crossing itself would be let as a design and construct contract and two commercial options were considered: either a conventional design and construct contract paid for directly out of Government funds or a privately financed scheme, whereby the successful consortium would design and construct the crossing, take over the debt on the old bridge and operate and maintain both crossings for the duration of the concession period.

For the design of the crossing, the contractors engaged Sir William Halcrow & Partners and SEEE of France as engineering consultants, and the Percy Thomas Partnership as architectural consultants for the tender bid. They also formed an alliance with the Bank of America and Barclays de Zoete Wedd to arrange the necessary financial backing for the possible concession contract. The tentative joint venture was one of four groups which successfully prequalified for the opportunity to bid for the contract.

Preparation of the bid

On receipt of the tender documents, a multi-disciplined, multinational tendering team was set up to prepare the bid.

The designers reviewed the Government's outline scheme and then worked closely with the contractors to develop the design of the structures and the associated construction methods concurrently. Drawings and other data were produced for the tender submission and to enable the detailed pricing of the scheme before the commercial arrangements could be completed.

Tender design

The design of the crossing submitted at the time of tender resembled the Government's outline design as far as the basic geometry of the structures was concerned. The horizontal alignment was identical and the overall layout was similar. This resulted in the main bridge over the Shoots navigation channel having a 456 m main span with sidespans of 99 m and the approximately 2 km long viaducts on each side having spans of 66 m or 99 m with a larger 132 m span over an existing British Rail tunnel.

The main bridge was designed as a cable-stayed structure, using in situ concrete pylons with a composite steel concrete deck. The viaducts were designed as a pair of precast concrete box girders to be erected using glued segmental balanced cantilever techniques.

The foundations for both structures were designed generally as spread foundations, constructed using precast concrete caissons. However, bored piles were used to ensure that the tunnel was not affected by the bridge construction and in those areas adjacent to each shore where significant depths of alluvium overlay the rock strata.

The design of the structure was carried out in accordance with the Government's firm requirements which gave a comprehensive design brief for the crossing. Although the structures were generally designed to BS 5400, the firm requirements called for significant enhancements for durability, wind loading and many other items. It also included requirements for ship impact, wind-shielding, maintenance and inspection gantries, wind tunnel testing, use of external unbonded prestressing tendons, protection of the British

PRE-CONSTRUCTION PERIOD AND DESIGN DEVELOPMENT

Proc. Instn Civ. Engrs, Civ. Engng, Second Severn Crossing, 1997, 13-21

Paper 11440

Neil Kitchener, chief engineer at Laing-GTM

David Mizon, a design manager at Halcrow-SEEE

Rail tunnel, cable stay removal, and so on.

The nature of the estuary itself was one of the most significant factors in the choice of design and construction methods. The tidal range at the site is 14·5 m at spring tides, which is the second highest in the world, and currents can exceed 5 m/s. At high tide, the river is over 4 km wide but at low tide, bedrock is exposed over the majority of the river, leaving the main channel some 350 m wide and a 500 m wide area to the west side as the only wet sections. The Bristol Channel is also seriously affected by the prevailing westerly winds and as well as frequent periods of treacherous weather, tides up to 2 m above predictions are not uncommon.

The regular exposure of the bed of the river effectively ruled out the use of conventional floating craft for all but light duties and all boats had to be capable of grounding safely at low tide, albeit on prepared foundations. The majority of the work was thus carried out by jack-up barges which could be manoeuvred safely at high tide and be available for use as a fixed construction platform at any state of the tide.

The nature of the tides was such that it was only possible to use floating craft for a few hours either side of high tide depending on the draught of the vessel and to use the dry river bed for access a few hours either side of low tide. This meant that there was at least a 2 h period every 6 h when access across the river was impossible, seriously disrupting patterns of work.

Award of the concession

After a six-month tender period, the tenders for both options were submitted in October 1989 and after a lengthy period of evaluation and negotiation the tender for the design, build, finance and operate (DBFO) option was accepted in April 1990. The concession agreement was finally signed in October 1990.

The contract for the concession was awarded to a company called Severn River Crossing (SRC). It was set up for the sole purpose of operating the concession by the four main parties with Laing and GTM owning 35% each and the Bank of America and Barclays de Zoete Wedd 15% each.

The concession was based on a required cumulative real revenue form of contract which allowed the concession company to recover £976 million in 1989 values over a period of up to 30 years. The capital expenditure included the cost of constructing the new crossing and taking over the £122 million debt on the existing bridge. At the start of the concession £62 million was repaid and £60 million will be paid at the end of the concession and is treated as a deferred debt on which interest is paid over the entire period. Toll levels were set in the tender for the three classes of vehicles, which became embodied in the Severn Bridges Act. They increase annually at the rate of inflation except for the first four years when they increased by 6% in real terms. On the basis of the original traffic forecasts, the £976 million revenue from the tolls will have been collected over 23

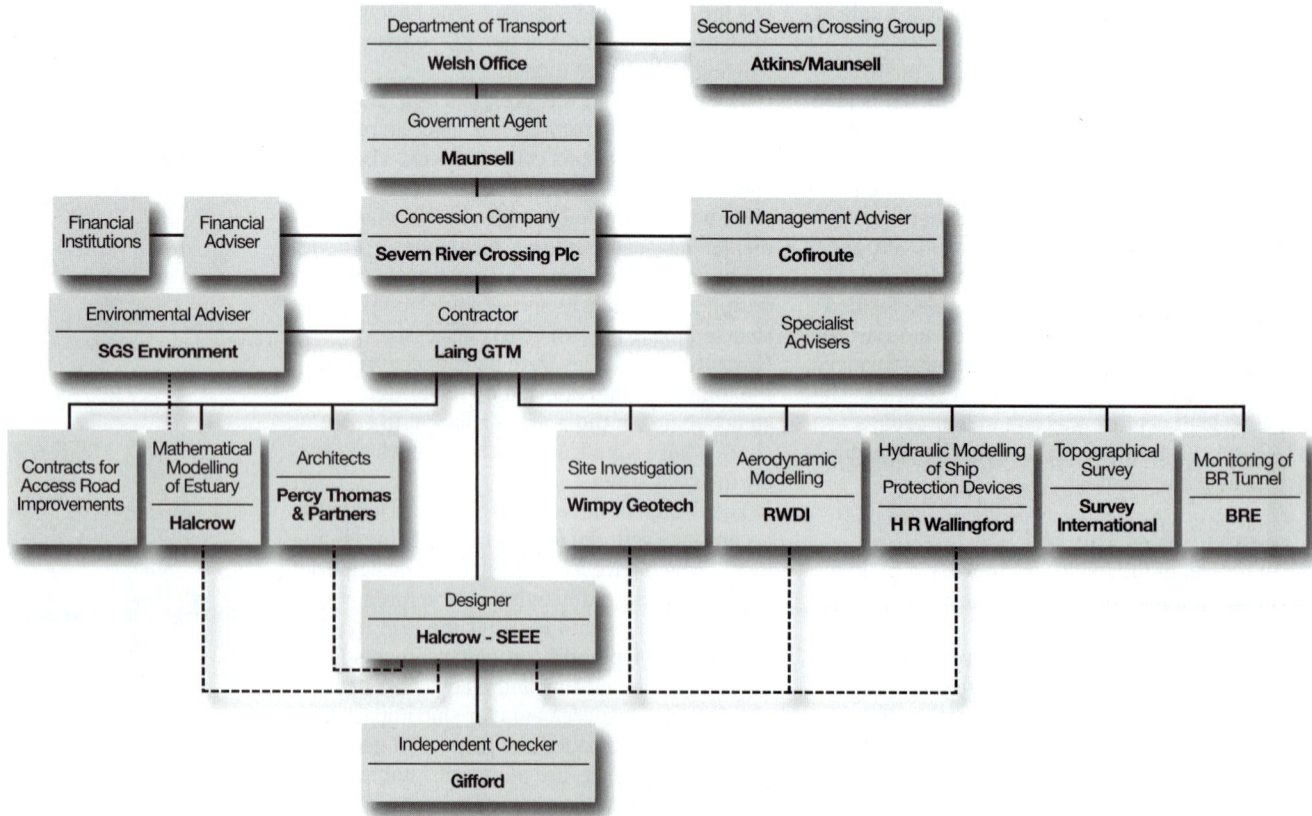

Fig. 1. Organization chart

PRE-CONSTRUCTION PERIOD AND DESIGN DEVELOPMENT

years and the concession will finish in 2015 when the crossings will be returned to the Government.

SRC then awarded contracts for its various areas of responsibility to in-house companies owned by its partners. A fully integrated joint venture between Laing and GTM was awarded the contract for the design and construction of the new crossing; the Bank of America and Barclays de Zoete Wedd were responsible for organizing the finance for the whole project; Cofiroute, a GTM subsidiary, was awarded the contract to manage the tolling operations for both crossings; and Laing Offshore was awarded the contract for managing the maintenance operations on both bridges. The design and construction contract was based on a modified form of the ICE 5th edition contract and was for a lump sum, with fluctuations based on the Baxter indices, of approximately £300 million.

Following the award of the contract, the two consultants, Halcrow and SEEE, were appointed as designers for the crossing and formed a joint venture Halcrow-SEEE. The joint venture established a project design team based in the Halcrow office in Swindon with work also being carried out in Cardiff, Paris and Lille. Fig. 1 illustrates the organization chart for the design phase.

Parliamentary process

The start date for the concession was dependent on the passing of the hybrid parliamentary bill which was required to enable the construction to start.

The design team worked on the development of the design, in conjunction with the contractor's team concurrently evaluating and preparing construction methods. Part of the parliamentary process involved public consultations and many displays and further public meetings were held to inform the public and those directly affected by the proposed works about the detail of the proposed construction measures and to describe the environmental protection proposals. Representatives from the contractors and designers were also required during the committee stages of the bill. In addition to the tender requirements, a number of other obligations arose out of these public consultations and the parliamentary process. These were all incorporated into the finally accepted version of the Government's firm requirements after the bill finished its passage through the House of Commons. The Severn Bridges Act 1992 received royal assent in February 1992 and the four-year construction contract started on 26 April 1992, the same day as the concession company took over responsibility for the existing crossing.

Philosophy of design and construction methods

It was established in the very early stages of preparation of the tender that if the project was to be successful, constant close cooperation between the contractor and designer throughout the design development period was imperative. Design work was not allowed to progress for any element unless the construction team were confident with their proposed construction methods for it. Hence design development both during the tender and after the award of the contract was based on numerous meetings between the contractor and designer punctuated by reviews by senior personnel on key features of the design.

During this process, the project architect was kept informed of all developments and attended many of the development meetings, thus permitting consideration of his views at each stage.

The early stages of the pre-tender design development period were spent establishing the fundamental construction philosophies and the main decisions made by the team during this phase remained virtually unchanged throughout the contract period. These are summarized below

- the form of the main bridge and the viaducts
- the length of the cable-stayed section of the crossing
- to minimize in situ construction
- to standardize all prefabricated items and size and weight limitations for each element type
- to the use of high strength friction grip (HSFG) bolts for all principal in situ steelwork connections with no site welding
- to develop the design of principal lifting and marine plant in parallel with permanent works design
- to design the foundations to withstand ship impact effects directly
- to investigate ways of eliminating or reducing the ship protection islands required by the contract at eight central pier locations
- to locate components of the bridge at spacings which suit the span module.

The particular features of the crossing site dictated many of the fundamental construction methods which strongly influenced the resulting design. The tidal range was the dominant feature. It was clear from the outset that operations in the estuary should be kept to a minimum and that precasting and prefabrication works were the key to successful design and construction.

Development of design and construction methods

Using the design philosophy described, the form and principal dimensions of key items were established in the early stages of the tender design programme. However, some preliminary investigations were carried out into different forms of construction. For the viaducts, steel plate girders with composite concrete deck slabs were examined, albeit briefly, since the contractor had a clear preference for concrete box girders. Various methods of box girder construction were considered and the chosen solution was precast units erected by launching gantry using glued segmental balanced cantilever techniques (Fig. 2). For the cable-stayed bridge, designs were examined using orthotropic and composite steel decks. The composite design with an in situ rather than a precast deck was adopted. Longer cable-stayed bridge options were briefly examined but it was clear that increasing the length of the cable-stayed bridge was less economic and would not enable the construction programme to be met. The chosen solution was a cable-stayed bridge of nearly 1 km overall length and a main span of 456 m, with foundations well clear of the side slopes of The Shoots channel (Fig. 3).

Although steel was the preferred principal material for the cable-stayed deck, fabricated steel pylons were rejected in favour of in situ concrete construction. The chief advantage of steel pylons is the rapid erection facilitated by large-capacity floating cranes. However, tidal conditions in the estuary ruled out the use of such equipment.

Initially, the viaducts were designed with spans of 99 m in the middle of the river and 66 m near both shores. In addition, the existence of the British Rail tunnel crossing the estuary at the

Fig. 2. View of the viaduct

Fig. 3. View of the cable-stayed bridge

eastern end necessitated piled foundations and a larger 132 m span, resulting in an untidy arrangement on the east side of the crossing. The shorter spans had constant-depth box girders, whereas the longer spans had varying-depth box girders with curved soffits. This scheme required 50 foundations in total.

One of the major decisions taken in the early stages of design development after award of contract was to standardize the viaduct spans to a single 98 m module requiring only 45 foundations. This enabled the viaduct erection gantry to be used for all spans without modification, reduced the number of piers, deck gantry moves and caisson placing operations and standardized the manufacture of deck units for every span. This change also improved the appearance and was welcomed by the Royal Fine Art Commission. It was found possible to adopt this standard span arrangement over the British Rail tunnel by designing special cantilevered pile caps to support each pier leg while maintaining the necessary clearance between the tunnel and the piles (Figs 4 and 5). Although the 66 m deck itself was cheaper than the 98 m deck the savings involved in reducing the number of foundations and standardizing the construction were significant.

The 98 m module had been extended into the backspans of the cable-stayed bridge in the tender design. Early pre-tender thinking was based on continuing viaduct construction up to each pylon but this was rejected since to meet the four-year construction programme, it was necessary to start construction of the cable-stayed bridge independently, without waiting for the viaducts to reach the pylons. However, the backspan pier spacing was retained at the viaduct module. This permitted the integrated design concept to be met and provided backspans of acceptable proportions. Using a balanced arrangement of cables each side of the pylons longitudinally raised the question of the need for the intermediate backspan piers. They are in a critical zone for ship impact and required special features to provide the horizontal shear resistance normally resisted by friction on each caisson base, owing to the reduced dead load effects of uplift resulting from live loads on the main span. A decision was taken to submit the tender with backspan piers, leaving the option of dispensing with them should detailed design justify this. However, in the early stages of post-tender design, their structural advantages were demonstrated despite their higher foundation costs and they were retained.

It could be argued that the backspan pier location was not the optimum, having been determined by the viaduct module chosen for a completely different structural form. However, the advantages in standardization of deck details and the overall appearance of the crossing were considered to outweigh those of locating them on the basis of pure structural optimization.

PRE-CONSTRUCTION PERIOD AND DESIGN DEVELOPMENT

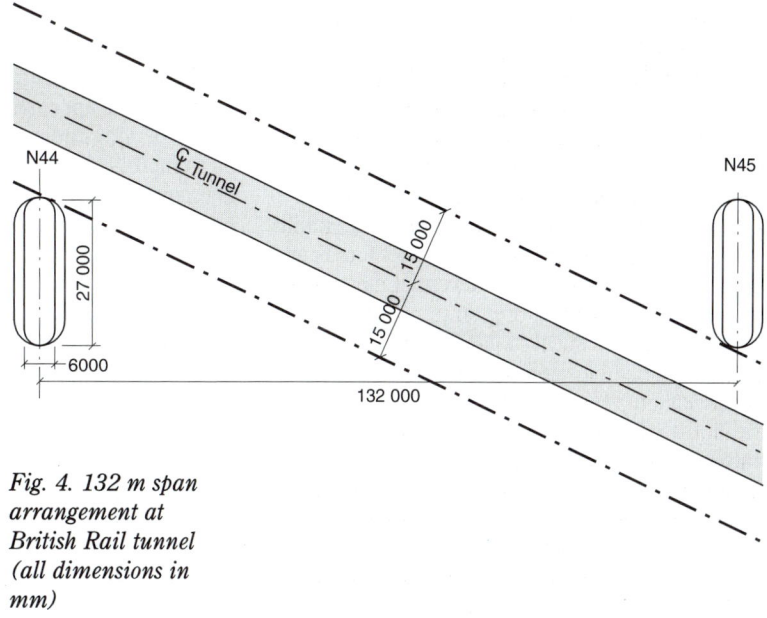

Fig. 4. 132 m span arrangement at British Rail tunnel (all dimensions in mm)

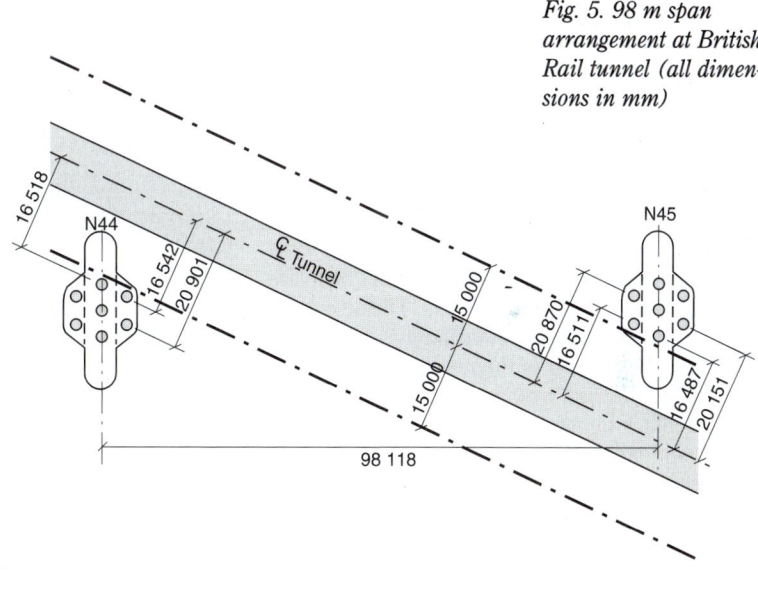

Fig. 5. 98 m span arrangement at British Rail tunnel (all dimensions in mm)

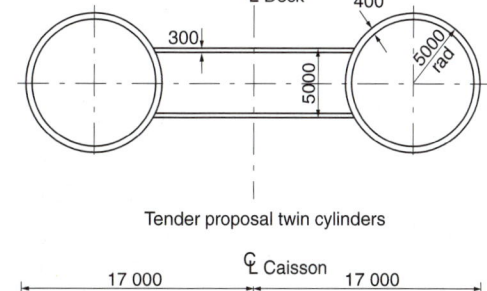

Tender proposal twin cylinders

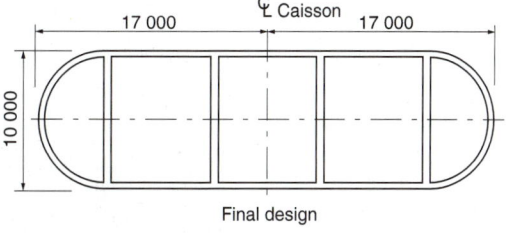

Final design

Fig. 6. Caisson options (all dimensions in mm)

17

A further significant development in the pre-tender design phase was the caisson shape. Initial concepts were based on twin circular precast units joined by precast walls, adopting the general theme of using precast shells in which the in situ reinforced concrete foundation was formed. These were abandoned in favour of the lozenge shape adopted, mainly because of difficulties with accommodating horizontal ship impact forces and anticipated problems with differential movement of the precast elements (Fig. 6).

Much time was spent in determining the optimum caisson shell sizes to provide the correct dimensions to resist ship impact, while keeping within the weight limit of 2000 t set by the contractor. Pre-tender decisions for this were based on simple calculations, owing to the limited time available at the tendering stage. In-depth dynamic analyses were not carried out until later for the detailed design. The level of the top of all the precast caisson shells was standardized at +8·0 m above Ordnance Datum, 0·5 m above maximum high water spring tide level, so that once dewatered, steel fixing and concreting of the internal works would not be affected by overtopping.

The tender design had in situ piers throughout but soon after the award of contract, a precast solution was developed to reduce the risks and plant requirement associated with in situ work on so many locations across the estuary. Having devised a system using precast hollow rectangular box units, various means of joining them were examined using high-stressed steel bars and various post-tensioning arrangements. The selected method used vertical post-tensioning.

A significant aspect of the development of the viaduct deck design was the introduction of intermediate blocking devices on the longitudinal prestress tendons to obviate the need for additional anchorages specified by a Government requirement. These are described in detail in other papers in this issue.

Various other features of the crossing evolved during the design phase and these are also described in other papers in this issue. However, once the basic configuration, principal features and dimensions of all elements of the crossing had been selected in the early stages of the post-award design phase, very little departure from these was found necessary and no significant changes to the design were made during construction.

During the design development the designer produced a detailed design manual called the 'design statement', containing the design philosophy, design loadings, method of analysis, standards, details of computer software, etc., for use by all design team members. This was a living document and was only finalized towards the end of construction.

In addition, the designer produced a two-volume technical specification for the project, together with all the working drawings, bar schedules and other design data necessary to construct the works.

Final design

In summary, the final design consists of a central cable-stayed bridge approximately 1 km long with viaducts approximately 2 km long on either side. The longitudinal arrangement is based on a 98 m module for the viaduct piers with all other elements spaced to suit. Shorter end spans were needed to suit local topographical features and structural economics. The vertical alignment is arranged to provide approximately 37 m clearance to mean high water level over the navigation channel. The double-curvature horizontal alignment follows that recommended by the Government and is illustrated in Fig. 7.

The final design incorporates a number of novel features necessitated by the particular conditions in the Severn estuary and the special requirements of the Government. In particular, the design and specification were prepared with the objective of producing a structure which meets the high durability standards necessitated by its location combined with a structurally economic and aesthetically pleasing bridge.

Fig. 7. Aerial view showing the double curvature of the bridge

PRE-CONSTRUCTION PERIOD AND DESIGN DEVELOPMENT

Fig. 8. Windshielding

A unique feature of the crossing is the provision of 3 m high windshields along the edges to protect motorists from the notoriously aggressive winds in the Bristol Channel. The windshielding significantly increased the wind loading on the structure and is illustrated in Fig. 8.

Investigations during the design and construction phases

To provide the contractor and designer with additional data on the physical characteristics of the site and to investigate certain aspects of the design as it developed, the following investigations were carried out on behalf of the concessionaire by subcontractors

- a detailed geotechnical investigation
- a hydraulic physical model testing programme for ship protection measures
- aerodynamic model tests
- a bathymetric survey of the estuary
- load testing of cable stays
- load tests on secondary cables.

In addition, Halcrow-SEEE carried out the following investigations and studies

- mathematical modelling of the hydraulics of the estuary

- examination of the feasibility of nosings to the central caissons instead of the ship protection islands required by the contract, making use of hydraulic physical model testing
- hydrogeological study of the effects of borehole and piling works on the Great Spring aquifer
- examination into the effects of likely ground movements on the British Rail tunnel crossing the estuary
- risk analysis on ship collision with the bridge caissons and piers.

The £5 million geotechnical investigation was one of the most extensive carried out in Britain and involved several marine craft and land-based equipment.

Although the tender design had been based on remote islands as ship protection, many alternatives to these were investigated, in particular rock nosings to the caissons. The physical requirements for these were examined by model tests in a specially constructed flume at Hydraulics Research, Wallingford.

A programme of sectional and aero-elastic model tests on the cable-stayed bridge was carried out in Canada, enabling the development of the deck design, particularly the edge detail, to achieve acceptable aerodynamic performance.

This is described by Irwin et al.[1]

To provide a model of the topography of the estuary bed, a bathymetric survey was carried out over the length of the crossing covering 1 km upstream and downstream.

The specification required that sample cable stays were tested for 2 million cycles of stress reversal and then tested to tensile failure. A series of tests was carried out in various testing establishments in Europe on test samples representing bridge cables with their corrosion protection sheathing and wax.

In addition the 'aiguilles' or secondary cables which restrain the cable stays from wind oscillations were subjected to fatigue testing with the clamps which attach them to the cable stays.

The mathematical hydraulic model was prepared principally to examine the effects on the hydraulics of the estuary of placing the large number of caissons and to examine likely scour. It was used additionally to provide data on currents for the planning of marine activities and for determining ship impact angles on the caissons.

The Great Spring is a deep aquifer which leaks high-quality water into the British Rail tunnel from where it is pumped out for commercial use. It was necessary to demonstrate that none of the ground engineering operations would cause any changes to the flow or quality of the water and this was done by a detailed hydrogeological study of the region. As a follow-up, Great Spring water quality was monitored daily during construction and no change in properties was recorded.

Following finite element modelling of the British Rail tunnel and surrounding ground with the adjacent bridge foundations, the resulting tunnel lining geometry changes were investigated by reference to the detailed construction of the tunnel.

Royal Fine Arts Commission approval

It was a Government firm requirement that each tender design was reviewed by the Royal Fine Art Commission (RFAC). Following this, the successful design had to be periodically reviewed by the RFAC during the design development period until they were satisfied that the design was visually acceptable and appropriate for its location in the Severn estuary. To assist with the RFAC presentations, a model was made for part of the cable-stayed bridge and a short section of viaducts. In addition, the design team architect produced a number of attractive watercolour paintings of the bridge and its environment.

Although the crossing design was favourably received right from the first pre-contract presentation, a few aspects of the design were identified by the Commission as needing improvement. These were

- pier widths
- east end span arrangement
- windshielding
- transition between the cable-stayed bridge and the viaducts.

As a result, significant attention was devoted to these areas during development of the design to achieve the required improvements. The most difficult of these was the junction between the cable-stayed bridge and viaducts, because of the fundamentally different structural forms. A satisfactory transition was eventually achieved by reducing the overall depth of the viaduct units at the junction and adding a profiled glass-reinforced plastic panel to achieve a similar external cross-section.

One other feature of the design considered vital to give the desired fully integrated appearance was the colour of the cables. The Commission insisted that the light green colour proposed for parts of the finished structure should be realized on the cable stays and this was achieved by adding adhesive tape of the precise colour specified by the architect to the black high-density polyethylene sheathing. The same colour was adopted for the windshielding, which, owing to its height, makes a significant impact on the bridge appearance and received much attention from the Commission.

Towards the end of construction a group from the RFAC visited the site and reported that they were satisfied that their requirements had been met.

Agreement with the designer

The pre-tender design was carried out under an informal joint venture arrangement between Halcrow and SEEE working for Laing-GTM on a timescale basis. Following award of contract, an agreement was signed uniting the two parties in a formal joint venture as designer for the project. This agreement covered both the design and the construction supervision phases. The basis of the agreement was that both parties were jointly and severally responsible for the design of the project and certification of the construction. Although the skills of Halcrow and SEEE were utilized to the best advantage of the joint venture, the work was carried out by an integrated, single team.

It was not until after award of contract that a formal agreement was signed between the contractor (Laing-GTM) and designer (Halcrow-SEEE). This was based on there being an agreement linking Halcrow and SEEE as one party and this concept was maintained throughout the whole project, with both the contractor and designer being considered as single parties. It was for this reason that the designer appointed one design manager and a designer's representative for the design and construction phases respectively to represent the joint venture. For continuity and convenience, the design manager took on the role as designer's representative when the construction stage started.

The design agreement between the contractor and designer was devised jointly, following the requirements of the construction contract which stipulated particular requirements of the design and its interface with the contractor and checker. The key aspects covered by this agreement were certification, programme, staff requirements, suspension, contractor's instructions, inspection and testing, changes, records, insurance, remuneration disputes and termination, for both the design and the construction phases.

The basis of payment by the contractor to the designer was a series of lump sums for both design and construction supervision phases. Similarly, the checker was paid on a lump sum basis. These lump sums were stated in the respective agreements.

The design agreement defined the end product of the design process as the design data. This included all drawings, reports, reinforcement schedules and so on needed to construct the crossing.

An important aspect of the agreement was design development, defined as 'that part of the design operation during which consultation takes place with or through the contractor leading to agreement of the criteria on which the final design will proceed'. This was found difficult to identify at times in the continuing exchange of ideas characteristic of a vibrant design and build atmosphere. This sometimes led to difficulties when reviewing the design progress as the point at which design development was completed and final design started was difficult to define.

The principal requirement of the scope of services was that the designer should enable the contractor to comply with the construction contract and to provide the design data in a timely manner such that the contractor was not delayed and could meet the works programme.

The designer's quality plan was based on the requirements of the concession agreement. It was submitted to the contractor for approval and was subjected to the Government's review procedure, and had to be endorsed as 'received' by the Government's agent before design work could start.

The design agreement required that the designer was reimbursed for additional works on a 'fair and reasonable' basis. This applied generally to design work requested by the contractor after construction had started, for example refining reinforcement layouts or changing details to suit revised construction methods. The designer's services terminated when the maintenance certificate was issued, with liability remaining 12 years after that date.

Agreement with the checker

It was a requirement of the construction contract that the design was independently checked and certified by a checker. This required working from final drawings and other design data produced by the designer, rather than checking calculations.

The designer was responsible for the checker's performance and for all certificates required from the checker. All matters concerning the checker's progress and other aspects concerning management of the checker were the designer's, not the contractor's, responsibility. However, the checker was paid directly by the contractor following approval of its invoices by the designer. Any additional works carried out by the checker were reimbursed by the contractor following the designer's approval. Any items of redesign carried out by the designer at the request of the contractor were subjected to the checking certification process.

Gifford & Partners was appointed as checker following submission of technical and commercial proposals requested from four selected firms of consultants, after the award of the main DBFO contract. The appointment was subject to signing an agreement with the designer (checker's agreement), which had to meet the approval of the contractor. This agreement was devised jointly by the checker and designer and followed the requirements of the construction contract. Some of the key aspects covered by this agreement were the programme, certification, rechecking procedure, payment, insurance, quality assurance, termination, arbitration and liability.

The checker appointed a project manager who was the first contact with the designer, there being no formal link between the checker and contractor. Since the checker was the last in line for releasing drawings for construction, the programme for submission of drawings, available checking time and dealing with any rechecks was critical to progress.

The checker's agreement required that the checker carried out its services such that the designer and the contractor could comply with the works programme and the design and certification procedure. The designer was required to submit design data to the checker 'at such times and in such forms that are reasonable in all the circumstances to enable the checker to comply with all its obligations'.

The checker was required to implement a quality assurance procedure complying with the quality plan prepared by the designer. This procedure was regularly audited by the designer. The Government's agent was permitted to audit both the designer and the checker against their quality assurance procedures.

During the construction phase the checker was not involved in the routine site supervision process but was informed of any changes required by the contractor and implemented by the designer as a result of the construction process.

Acknowledgements

The authors wish to acknowledge with thanks the assistance received from many of their colleagues in the preparation of this paper and in particular C. M. Tong of Halcrow-SEEE. Photographs are reproduced with kind permission of Photographic Engineering Services.

Reference

1. IRWIN P., MIZON D., MAURY Y. and SCHMITT J. History of the aerodynamic investigations for second Severn crossing. *Int. Conf. AIPC-FIP, Deauville, France*, 1994.

Second Severn crossing— design and construction of the foundations

J. N. Kitchener, BSc, CEng, FICE, *and S. J. Ellison*, BEng, CEng, MICE, MIHT

This paper describes the design and construction of the foundations for the second motorway bridge over the Severn estuary between England and Wales. The two main piers of the central cable-stayed bridge and the 49 supports of the approach viaducts are founded mostly on exposed sandstone and mudrock. Owing to strong currents and the high tidal range, the contractor chose to construct 37 of the foundations using precast caissons weighing up to 2000 t. These were floated out from the casting yard on the English shore and placed using specially adapted barges. The remaining foundations were piled through alluvium or to span an existing rail tunnel. All foundations were designed for potential ship impacts.

Neil Kitchener, chief engineer at Laing-GTM

Stephen Ellison, substructure design team leader at Halcrow-SEEE

Design

Ground conditions

Bedrock exists at the surface over much of the area of the second Severn motorway crossing between England and Wales and comprises Triassic sandstones and mudrocks. The strengths of the sandstones and mudrocks vary, with uniaxial compressive strengths between 7·5 MPa and 15 MPa. On the Gwent side in Wales, 6–10 m of soft alluvial deposits overlie the bedrock for a distance of 1 km. On the Avon side between piers N37 and N46, a layer of mudrock is exposed at the surface and is underlain by weaker Keuper Marl. The thickness of the sandstone layer varies between 0·5 and 4 m. Fig. 1 shows the geology of the site.

Ground investigation

Although some site investigation data had been provided at the time of tender, an additional site investigation contract was carried out in 1990/1991 with boreholes being drilled mainly from jack-up barges. In general the boreholes were located at the anticipated foundation posi-

Fig. 1. Geological section

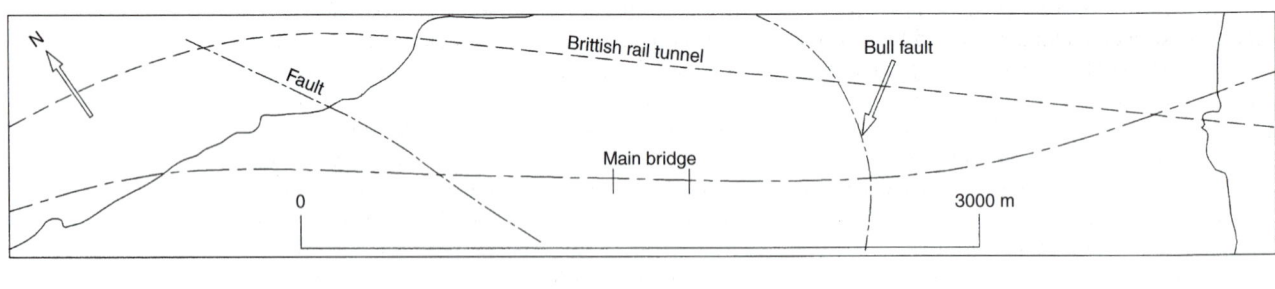

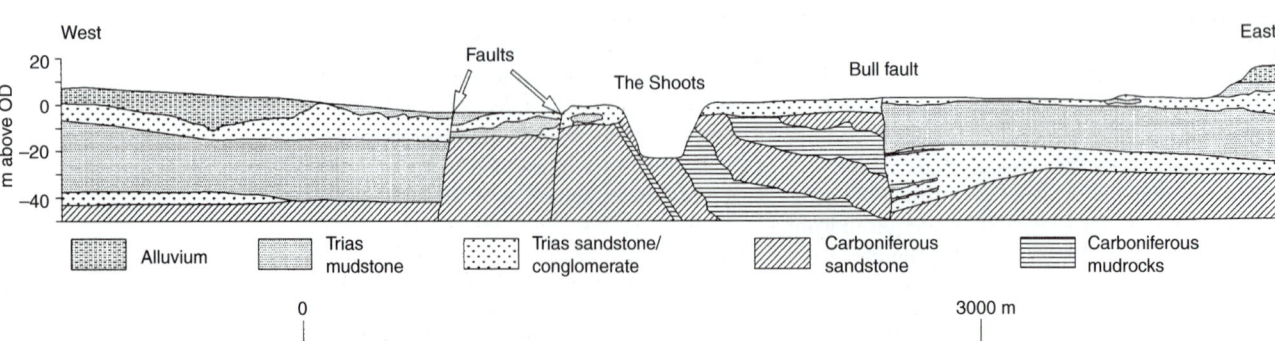

DESIGN AND CONSTRUCTION OF FOUNDATIONS

tions. Plate bearing tests, impression packer tests, packer permeability tests and dilatometer tests were carried out in some boreholes. In addition a programme of geophysics testing was carried out, comprising borehole geologging, cross-hole shooting, marine seismic reflection and refraction, land seismic refraction, echo-sounding and side-scan sonar.

In addition to the offshore work, a large-scale in situ shear test was devised to estimate the shear strength of the sandstone. It was recognized at an early stage that horizontal ship impact loads would be a critical load case for the design of the foundations. The frictional shear resistance between the concrete foundation and exposed bedrock and the shear strength below this level were therefore important parameters. Shear tests were undertaken in which lateral and vertical loads were applied to an in situ block of undisturbed rock. These results, together with the results of laboratory tests on borehole samples, provided a basis for establishing the shear strength of the rocks, with ø values taken as 30° for sandstone and 26° for mudrock.

Data from the ground investigation was used for the design of the caisson and piled foundations, temporary works foundations, settlement analyses, scour assessments, hydrogeological investigations, etc. Details of the ground investigation are given by Chambers et al.[1] and Madison et al.[2]

Selection of foundation types

Several factors influenced the choice of foundations. Piled foundations were adopted for areas where alluvial deposits overlie bedrock. The pile caps were constructed either in situ or using a precast concrete shell, depending on accessibility. Special piled foundations were also required adjacent to a British Rail tunnel which crossed the site. The existence of sound rock at or near surface level over the remainder of the crossing meant that spread foundations would be appropriate without the need for widespread piling. In addition to the loading from the bridge superstructure, a significant lateral load due to potential ship impacts needed to be considered and this further reinforced the preference for large spread foundations with high sliding resistance.

The most significant feature of the estuary is the extreme tidal range. Consequently much of the estuary bed is exposed at low tide, which permitted possession of the foundation locations for a few hours per day. However, associated with the high tidal range are high flow velocities. While access to the formation level over much of the site was seen as a generally positive aspect, the high stream-flow velocities imposed a substantial constraint on the method of construction and, in particular, the interim construction stages, which needed to be sufficiently robust to withstand water currents of up to 5 m/s twice a day. Temporary formwork was considered to be too light and its reuse would be very limited after exposure to sea water. Precast concrete caissons, which had the advantage of offering a permanent formwork shell to the infill core and at the same time offering a large single-component element which could be placed by a relatively fast operation within the low-tide window, were therefore selected. Open-bottom caissons were preferred to closed-bottom as the latter would need a gravel or sand bedding layer, which would be extremely difficult to place in such high currents and be liable to wash out after caisson installation.

Four different sizes of caisson were adopted as illustrated in Fig. 2 and Table 1. The principal determining factors in the sizing of the caissons were the magnitudes of vertical and lateral loads caused by ship impact and various combinations of these.

Caisson design

Caisson design consisted of two distinguishable parts, the global design as a spread foundation and the local design of the component parts.

The global design of the caissons covered the normal checks for a spread foundation, including permissible bearing pressures, overturning and sliding. The critical load effect varied, although in general, ship impact was the determining load for the sizing of the larger caissons, which were positioned closer to the navigation channel.

A basic set of guidelines for the application of

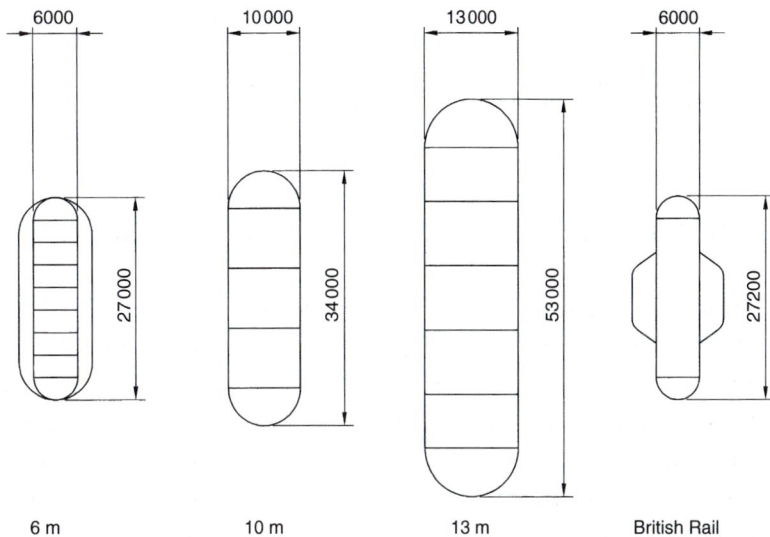

Fig. 2. Caisson sizes (plan on caisson shells; dimensions in mm)

Reference number	Location	Type
N1	Gwent abutment	Abutment—piled
N2 to N11	Gwent viaduct	Piled
N12 to N22	Gwent viaduct	10 m caisson
M1, M2, N23 to N26	Cable-stayed bridge	13 m caisson
N27 to N36	Avon viaduct	10 m caisson
N37 to N43	Avon viaduct	6 m caisson
N44 to N45 (British Rail tunnel)	Avon viaduct	Special piled caisson
N46	Avon viaduct	6 m caisson
N47 to N48	Avon viaduct	Piled
N49	Avon abutment	Abutment—piled

Table 1. Schedule of foundations

the ship impact force was given in the Government's requirements[3] but it was felt that for such an important load case, the criteria needed to be developed further to provide a coherent and appropriate basis for taking account of all the significant variables economically.

Ships of different dead-weight tonnages ranging from 200 to 6500 t were considered, depending on the proximity to the main navigation channel. The change point from 10 m to 6 m caissons coincided with the zoning of ship sizes. The basic impact load was derived using the formula given in the Government's requirements and expressed as

$$F_{impact} \text{ (MN)} = 0.44 \sqrt{(dwt)}$$

where *dwt* is the dead-weight tonnage of the ship.

This formula takes no account of vessel speed and it was recognized that while this is an acceptable assumption for ships travelling at a relatively high speed, it is conservative for lower speeds, for instance in the case of drifting ships. With assistance from specialist marine engineering consultants, a study of the force build-up and vessel crushing characteristics was undertaken, from which a set of curves was developed to estimate the reduction factor for different vessel speeds.

Further reduction factors were introduced to take account of the direction of travel and the degree of penetration. Using the designer's mathematical hydraulic model of the estuary it was possible to create a full set of impact scenarios taking into account vessel size, whether it was in ballast or laden and the tidal cycle.

The load factor for the ultimate limit state was taken as 1·5 for all ship impact loads. However, a risk analysis taking account of ship sizes and movements together with tidal cycles was undertaken, which demonstrated that this factor could be reduced to 1·0 for the case of ships under power colliding with those viaduct foundations located away from the main shipping channel. This led to savings for some foundations.

A finite element ground model was developed to estimate vertical and horizontal displacements for the critical foundations and to assist in determining satisfactory founding levels. Typical bearing pressures ranged between 600 kN/m² and 1800 kN/m². Suitable formation conditions were present at most locations either at or very close to the bed level. Minimal preparation of the formation was specified and in most cases this amounted to trimming up to 1 m from the surface rock. Typical settlements of 30 mm to 50 mm were estimated and the actual settlements recorded were generally less than 25 mm.

The design of the foundations between piers N37 and N46 was more problematic. The Triassic sandstone in this area is less than 1 m thick in some locations and overlies softer Keuper Marl. It was estimated that the permissible bearing pressure on the Keuper Marl was in the region of 600 kN/m². The 6 m caisson which was developed for the smaller-ship impact zones was modified by the addition of a foot or extension at the base which effectively increased the width to 10 m. As a further measure, the thickness of the rock layer was effectively increased by the addition of a mass concrete platform where the sandstone was particularly thin, thus helping to spread the stresses over a larger area of the underlying Keuper Marl.

Finite element analysis was also used in the analysis of ship impact, although in this case the dynamic effect was an important additional consideration. A quasi-static analysis was undertaken with a dynamic enhancement factor (DEF) applied to the static load.

Initial analyses were made using a three-dimensional dynamic model but this was abandoned because of its complexity. Instead, several two-dimensional finite element models of the caisson and ground were created which allowed a study of the various parameters. The first set of models represented a vertical plane intersecting the caisson–ground interface for various scenarios. A DEF of about 1·2 was established from the results.

The quasi-static impact force modified with a suitable DEF was then applied to a second set of finite element models, this time representing the horizontal plane between the caisson and the rock formation. These models incorporated a sliding interface based on the interface friction obtained from the site investigation. The model was able to determine the overall sliding resistance of particular caissons subject to eccentrically applied impact loads causing twisting, as well as calculating the vertical stresses imposed during impact. Each of the finite element models used for the ship impact analysis was developed using ABAQUS.[4]

All of the caisson foundations were designed as simple spread foundations, except for the backspan caissons of the cable-stayed bridge. Positive loading on the main span of the cable-stayed structure results in uplift forces on the backspan piers and the total vertical load on these foundations is much less than on similar caissons supporting the viaduct. From the finite element models, it was possible to ascertain that resistance to sliding would have to be provided by means other than self-weight. Initially, a combination of ground anchors and shear keys was designed to overcome the related problems of sliding and overturning. However, concerns over the long-term durability of the ground anchors led to the decision to increase the size of the caisson from 10 m to 13 m wide, which added sufficient weight to prevent overturning, together with shear keys in the form of 2 m dia. tubular steel piles cast into 2 m deep holes excavated below the general foundation level to resist sliding.

The structural form of the caisson foundation comprised two main components, the precast shell and the in situ concrete core. The design of the shell was dominated by the load effects induced during construction and in this respect, the critical caissons were the largest, namely those for the pylons. The lifting capacity of the construction equipment limited the weight of all shells to 2000 t. With overall dimensions of 53 m x 13 m x 11 m tall, this presented a serious constraint on the design of the pylon caissons. The weight was limited by adopting a relatively thin wall 400 mm thick together with internal steel bracing to provide both torsional strength and lateral stability.

The shell structure was subjected to a range of load effects both during transportation and when in place. The designer's detailed mathematical model of the estuary provided data on current speed and direction which was invaluable in the design and construction planning of the foundations. The dynamic effects of lifting, together with accelerations due to rolling and pitching while being transported by the barge, were carefully examined and a load enhancement factor included where appropriate.

The caisson was lifted at four points, with one pair of lifting points linked by a common hydraulic jack mechanism to create a standard three-point lifting system. Owing to the inherent torsional flexibility of the shell structure, this meant that high local stresses near the supporting points arose from imposed lateral loads during transportation. A finite element model was used to calculate the stresses at each stage and was particularly useful in quantifying the high local stresses around the lifting points.

Once in position, the caisson was supported on a series of grout-filled bags or sacks which were inflated between the underside of the shell and the formation. This method of support ensured an even distribution of load. After placing, both stream-flow and wave forces became significant. It was imperative to stabilize the caisson shell quickly by pouring a mass concrete plug to fill the lower region, which was cast to the minimum height necessary to prevent flotation after dewatering. Concrete cross-walls were incorporated in the lower part of the shell to provide restraint against the stresses induced by this mass concrete and improve rigidity during transporting. Even so, heavy longitudinal reinforcement in the shell was still needed to limit crack widths to 0·10 mm. The cross-wall height was limited to the depth of the plug because of the weight restriction on the shell and also to maintain an unrestricted area for the construction of the reinforced base to the pier or pylon above.

Once the plug had gained adequate strength and the caisson was dewatered, construction of the reinforced portion of the base was carried out in controlled conditions. The formation of the upper part of the core caused high stresses in the shell wall due to both the hydrostatic effects of the wet concrete and the thermal expansion from hydration. The latter was particularly onerous since the restraining effect of the steel bracing was largely diminished as the steel also expanded with the increase in temperature of the surrounding concrete. The stresses in the shell wall were predicted by estimating the temperature rise of the infill concrete and then considering strain compatibility between the shell and the core. The temperature rise was calculated from in-house software and was found to compare very well with actual field measurements. This provided the basis for detailing the reinforcement in the shell walls and the sequence and rate of pouring of the infill concrete to minimize the build-up of stresses. A typical pour sequence is shown in Fig. 3.

For global behaviour, the reinforced portion of the infill concrete was designed to be integral with the shell and the transfer of shear stress between the shell wall and the infill was achieved by shear keys cast on the inside faces of the shell wall.

Piled foundations

The piled foundations were located in zones where ship impact loads were small. Nevertheless, ship impact was still a critical load condition and it was originally intended to use 1·5 m dia. raking piles to provide the necessary lateral resistance and the tender was based on this. However, the installation of raking piles in the estuary was subsequently rejected owing to the problems associated with boring raking piles and 2 m dia. vertical, bored, cast in situ piles with more reinforcement were used instead, with little additional cost.

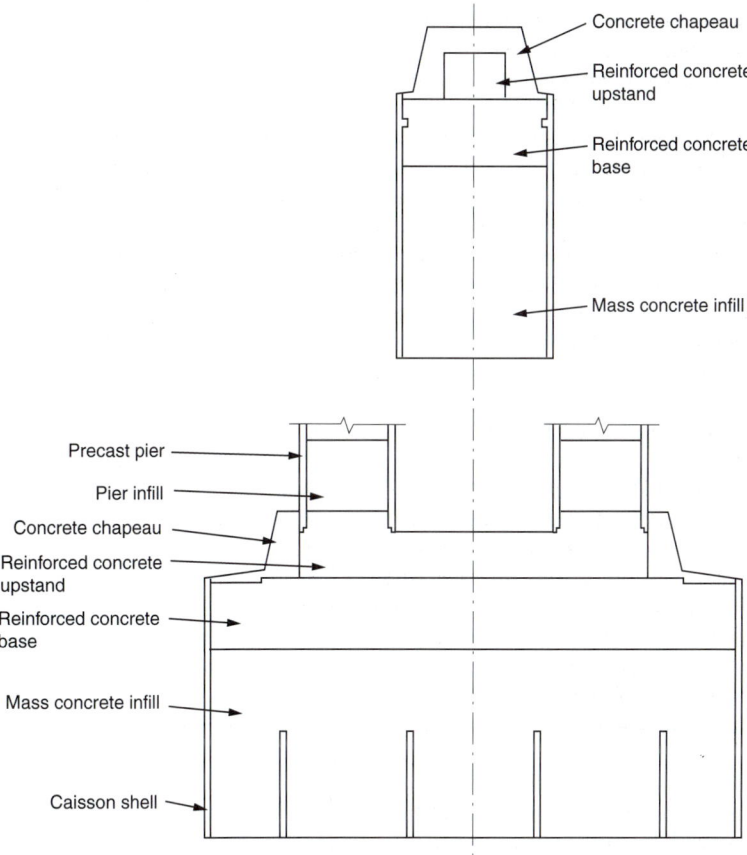

Fig. 3. Caisson—typical pour sequence

The assessment of ship impact loads followed the same principles as for the caissons although the magnitude of the horizontal load in this case was only about 6 MN, compared with 25 MN for the caissons. The DEF was established using a non-linear analysis of the pile–ground interface using LARSA[5] and was found to be close to unity. The upper portion of the pile cap was shaped to improve impact angles and hence reduce the torsional effects of the ship impact load.

MPILE[6] was used to calculate the vertical and lateral loads in the piles. The vertical working load of the piles was typically 13–17 MN and the pile lengths were generally a little under 20 m. Each pile incorporated a permanent steel casing which was not used in the strength calculations. A 2 m dia. trial pile near pier N3 was tested to twice its working load. Two further trial piles 1 m in diameter were constructed near to each abutment and also tested to twice their working load. In addition, a lateral load test was carried out on each of the 1 m piles to verify the assumptions made for the ground parameters used in the calculations. The test results correlated well with the calculated pile capacities. Further selected piles were load tested to 1·5 times their working load as works test piles.

A temporary earth bund supplemented with sheet piling was used adjacent to the Gwent shore to create a cofferdam for the construction of the piled foundations N1 to N8. Conventional in situ pile caps were used for these foundations.

The piled foundations lying outside the cofferdam incorporated a precast shell acting as permanent formwork to the reinforced infill. Similar principles to the caissons applied, except that in this case the shell was supported directly from special brackets attached to the pile heads and although similar in form to the caissons, these were in effect large precast pile caps.

The abutments at the Gwent and Avon shores were conventional reinforced inverted 'T' forms. Bored, cast in situ piles 1 m in diameter were used. In the case of the Gwent abutment the approach embankment behind the abutment was founded on compressible strata, which gave rise to lateral loading on the piles caused by squeezing of the soft strata. Additional longitudinal reinforcement in the piles was designed to resist the resulting imposed bending moments.

British Rail tunnel foundation

Some 200 m from the Avon shore, the crossing alignment passes over the British Rail tunnel carrying the main rail connection between Wales and England. The tunnel is over 100 years old and has both historic and functional value. British Rail was understandably concerned about surcharge loads and the installation of piles in the vicinity of the tunnel and as a result it was stipulated that no foundation should be within 15 m of the tunnel.

The span arrangement conceived during the tender period was based on module spans of 66 m in this region with a span of 132 m over the tunnel in order to comply with the 15 m exclusion zone. As the viaduct design developed, the module spans were changed to 98 m. In order to maintain the standard pier arrangement and foundation appearance, a special pile cap to support the viaduct piers was designed which effectively cantilevered out from a group of seven 2 m dia. piles concentrated at the centre of the foundation (Fig. 4).

The piles were debonded from the ground throughout their length so that axial load was resisted only by rock-socketing in the bedrock below the tunnel invert level (see Fig. 5). Although the piles were located outside the 15 m exclusion zone, it was nevertheless necessary to demonstrate that no adverse stresses would be transferred through the surrounding ground to the tunnel lining. A two-dimensional finite element model was created to represent the tunnel, ground and piles. Typical stress outputs are illustrated in Fig. 6. The analysis predicted deformations in the brick-lined tunnel of less than 6 mm,

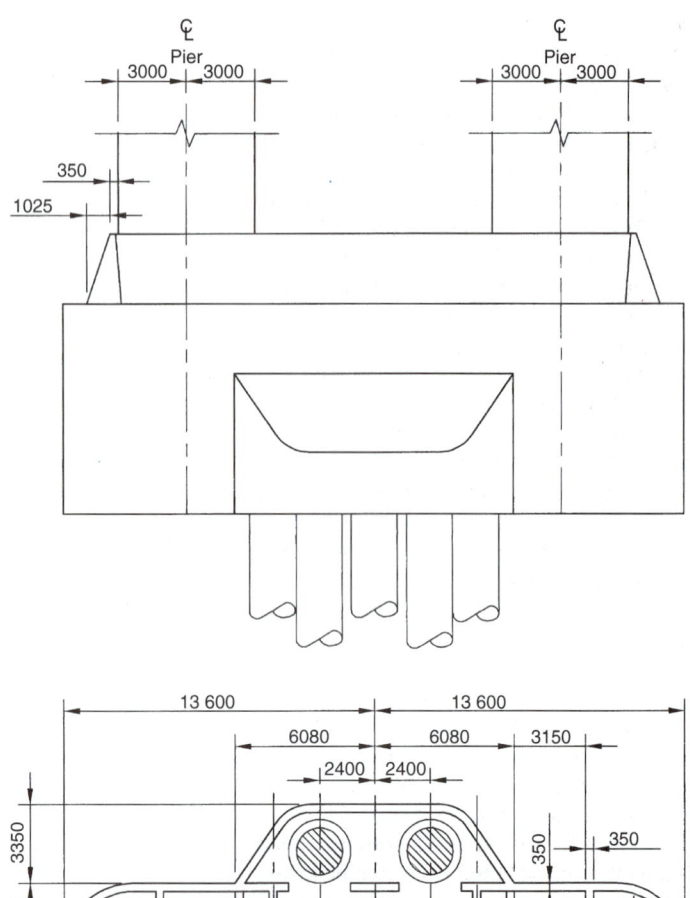

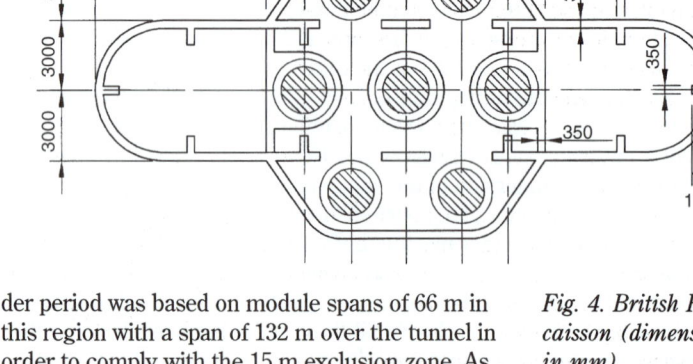

Fig. 4. British Rail caisson (dimensions in mm)

which were considered acceptable.

Sections of the tunnel in the area of the works were instrumented to check on the incidence of movements of the lining. These showed that by far the greatest movement was caused by the tide.

The pile cap comprised a precast shell with in situ reinforced concrete infill. Although lifted by the same crane as for the main caissons, the weight was limited to much less than 2000 t because the crane had to be operated with an extended jib in order for the jack-up legs to be positioned outside the tunnel exclusion zone. The bottom of the pile cap shell was closed and after sealing around the piles it became necessary to hold down the caisson onto the piles with temporary prestressing to overcome buoyancy at high tide. Once sufficient infill concrete was placed, the holding-down prestress was released and the in situ concrete core completed.

Protection to ships

Ship impact concerns were not restricted to minimizing damage to the bridge structure but also considered how to mitigate damage to the vessel itself. Apart from the cost of repairs or replacement of the vessel, the potential loss of human life and environmental damage were very real concerns. Therefore, it had originally been envisaged that artificial islands would be constructed at or close to some of the caissons at strategically chosen locations to prevent a vessel being seriously damaged in the event of a collision with the bridge, even though the bridge itself was designed to withstand such collisions. The islands were to be constructed from graded rock in order to absorb much of the energy of a vessel collision.

During the detailed design of the substructure, it became possible to extend the risk analysis initiated for the study of impact loads on the caissons to investigate the levels of risk which existed for the vessel owner, crew and harbour authority, Gloucester Harbour Trustees. A full quantitative risk analysis was undertaken which showed a very small risk to vessels, thus enabling a decision to be made not to use artificial islands but to provide instead some additional navigation control aids to Gloucester Harbour Trustees.

Foundation construction

The tender concept for the construction of the foundations was to manufacture the caisson shells on the shore, transport them down a slipway into the river bed at low tide and pick them up by a self-propelled jack-up catamaran which would then transport and place the unit in position on the river bed.

There was no equipment suitable for this particular operation, so tenders for the design and supply of a suitable vessel were invited. The lowest bid received, at £23 million, was over twice the budget and with the extremely limited resale

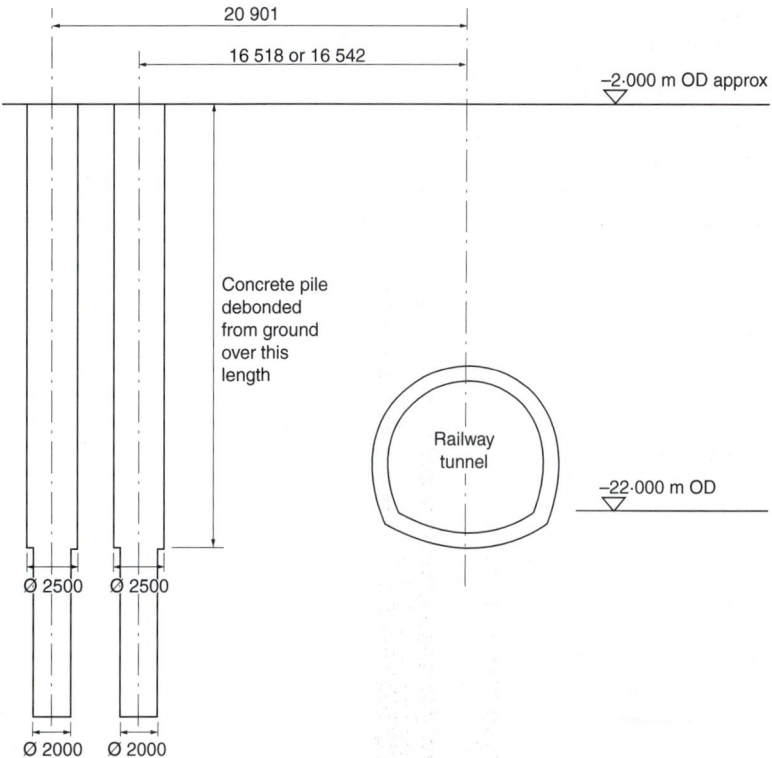

Fig. 5. Piles for British Rail caisson (dimensions in mm)

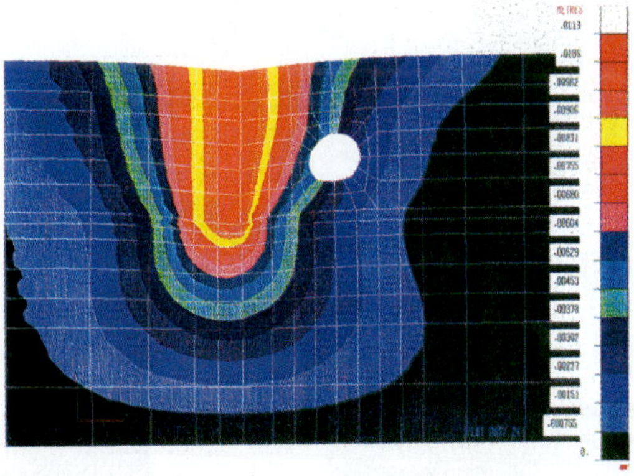

Fig. 6. Typical stress outputs for British Rail caisson foundations

potential, other methods were investigated.

The method finally chosen involved converting a dumb flat-bottomed barge, the *Sar 3*, into a powered vessel by adding four thruster propulsion units to transport the units across the river at high tide and converting an existing four-legged jack-up barge, the *Lisa A*, into a crane platform to support two 1250 t Lampson Transilift cranes for placing the units. The overall cost of converting both vessels was also over budget but their versatility meant that they could be used for many other operations, particularly on the cable-stayed bridge, resulting in considerable savings in these operations.

The design of the caissons and the handling equipment was carried out concurrently but in order to enable the design of the equipment to be finalized, the weight of the caisson was frozen at 2000 t some 18 months before the design was completed. However, as a result of design developments and changes to the construction, it proved difficult to restrict the weight and it would have been preferable to have been able to handle units weighing up to 2200 t.

Marine access and temporary works

Ground conditions and access were much better on the English side of the river and it was decided to concentrate construction activities on this bank. A significant amount of preparatory and temporary work was carried out to prepare the construction yard.

Reinforced earth was used to form a large landing jetty and ro-ro ramp for loading the *Sar 3*, which berthed against tubular steel mooring dolphins piled into the river. At low tide, the barge, which had been suitably reinforced, rested on a specially prepared concrete/timber foundation which simplified the loading of the vessel. A series of floating pontoons, accessed by a bridge, were used as berths for the passenger, safety and work boats. A tubular steel loading jetty for all the marine concreting materials was constructed along with the associated silos, pumps and con-

Fig. 7. Causeway 1·5 h before low tide showing partially constructed caissons

veyor systems for loading the materials barge.

A reinforced-earth access ramp to provide vehicular access to the river bed was also constructed using no-fines concrete blocks and Paraweb.

In order to gain access across the English Stones at low tide, a 2 km long causeway was constructed from the shore out to the east pylon of the main bridge. For the majority of the route, the construction was restricted to clearing loose material and concreting a slab using dry, structural-grade concrete compacted by roller. It was possible to do this in most conditions except during high spring tides, when serious erosion occurred.

Over a 500 m section of the route, the bed of the river was up to 2·5 m lower than the general area and it was necessary to construct a deeper causeway to the same level in order to maximize the length of time over which it was available for use. This was achieved using a combination of rock bunds and precast concrete culvert units which allowed the water to ebb and flood at low tide without overtopping the roadway (Fig. 7).

Foundation preparation

On the Gwent side, a protective earth bund (Fig. 8) was constructed out from the shore and the large-diameter bored piles for the first eight foundations were constructed behind this protection using conventional equipment. Sheet-piled cofferdams were also used to enable the concrete pile caps for the last two foundations to be constructed in dry conditions.

The 2 m dia. piles for the next three Gwent foundations and also for the two which straddled the British Rail tunnel were drilled using a reverse circulation rig mounted on a jack-up barge. The concrete was placed by pumping from truck mixers using the causeway or earth bund as access. On the Avon side the bored piles for the first three foundations were constructed on shore using conventional equipment.

The foundations for the remaining bases were formed using caissons, sitting on a prepared formation which varied depending on the particular location. The formation level of each caisson was determined from the site investigation and for those founded on the thick layers of sandstone, any deleterious material was removed and the rock was excavated to a constant level across the whole area. This level was then compared with the proposed level and the caisson height was adjusted if necessary in the precast yard to ensure that the top level remained constant.

The excavation plant tracked out along the causeway at low tide for those foundations near to the shore. For those near the middle of the river and on the Welsh side of the navigation channel, the plant was carried on a converted coaster which floated at high tide but beached in a suitable location at low tide. This allowed the machinery to access the river bed using a ramp lowered over the side of the vessel.

Fig. 8. Gwent bund

For those caissons founded on concrete pads, minimal preparation of the rock to remove soft or loose material was carried out. The first pads were constructed in a number of pours using steel formwork and conventional wet concrete poured at low tide. However, a large number of the pours suffered from wash-out of the concrete when the tide came in and the formwork was also susceptible to damage and displacement. The later pads were successfully constructed using the roller-compacted concrete technique which had been perfected on the causeway.

There were six foundations on the Welsh side which were permanently underwater and the six-legged jack-up barge *Jay Robertson*, which had been modified to carry a 200 t Manitowoc 4100 ringer crane, was used to position a heavy structural steel platform on the bed of the river. At low tide, an excavator was lowered onto the platform to excavate the soft rock above the founding level into barges for transportation to the tipping area. The frame was moved back along the base as the excavation proceeded. On completion, the resulting hole was backfilled with concrete which was placed with tremie tubes during periods of neap tides in order to reduce wash-out damage.

The steel piles used to increase the shear capacity of the backspan piers for the cable-stayed bridge were installed in holes excavated into the rock in the centre of the foundation and concreted into position.

Caisson construction

Initially, six piled casting beds for caisson construction were built in the Avon construction yard but a further two were added six months into the contract in order to maintain the programme (Fig. 9). The thin-walled shells weighing up to 2000 t were constructed on the casting bed using conventional construction techniques, plywood/aluminium formwork, scaffolding and tower cranes in up to 30 individual pours. The complicated grout bag pods at the intersection of the cross-walls and the outer walls were pre-cast on site and incorporated into the in situ walls. On completion, the caissons were fitted out with the internal structural steel bracing, which provided temporary stability during lifting, transporting and concreting operations, as well as the drain pipes and valves used to control flotation.

Caisson lifting and transportation

When a suitable weather and tide window were forecast, the caisson was lifted 5 m above the ground using a system of four 600 t heavy-duty strand lifting jacks (Fig. 10). A pair of Lampson crawler transporters carrying a purpose-made steel frame were driven under the suspended caisson which was then lowered onto the spreader frame. The fitting out of the grout bags and pipes, temporary ladders, access platforms and survey targets was then completed prior to float-out.

At low tide, the Lampson crawlers transported the caisson around the site and down the ro-ro ramp onto the *Sar 3*, which was then ballasted to ensure that it subsequently floated on an even keel (Fig. 11).

Fig. 9. Avon casting yard showing caisson casting beds and a caisson being transported down ro-ro ramp

Fig. 10. Lifting 2000 t caisson for pylon foundation

DESIGN AND CONSTRUCTION OF FOUNDATIONS

Caisson placing

Prior to this, the *Lisa A* had been pushed into position at high tide by the *Sar 3*, which was designed to latch onto all the dumb barges and manoeuvre them around the estuary (Fig. 12). Because of the limited slewing capacity of the Lampson cranes, the *Lisa A* had to be located within 500 mm of its theoretical position with respect to each caisson. The initial proposal had been to place the *Lisa A* as near as possible to the target, run out winch lines to prepared anchors at low tide and then use the winches to move accurately into position at the next high tide. In the event, it was found that the *Sar 3* was capable of placing the *Lisa A* to the required degree of accuracy and the winches were rarely used.

Once the *Lisa A* was on station and jacked out of the water, the front flotation tanks were filled with a predetermined quantity of water in order to preload the legs into the river bed. This tested the footings and ensured that there would be no further settlement when the caisson was lifted. On completion, these tanks were discharged and the rear tanks filled with water to act as a counterweight for the lift. The barge was jacked up to its operating level some 10 m above high tide and the position accurately surveyed using a set of GPS units linked by cable to the on-board computer. The data from the shore-based reference station was transferred by a telemetry link in real time.

The float-out operation, which took up to three hours, was restricted to a three- or four-day win-

Fig. 11. *British Rail caisson being transported down ro-ro ramp*

Fig. 12. *Lisa A being manoeuvred into position by Sar 3*

Fig. 13. Sar 3 transporting a caisson across the estuary

Fig. 14. Lisa A in position ready to lift a caisson

DESIGN AND CONSTRUCTION OF FOUNDATIONS

Fig. 15. Lisa A lifting a caisson

Fig. 16. Lisa A waiting to lower a British Rail caisson at low tide

dow either side of the spring tides because there was insufficient water for the craft to navigate across the Stones at other periods.

On the appointed tide, the *Sar 3* set sail (Fig. 13). It approached the *Lisa A* and stood off some 20 m while its position was stabilized using the laser-operated dynamic positioning system (DPS) coupled to the computer control system for the thruster propulsion units. Once the position was stable, the DPS was used to move the *Sar 3* progressively towards the *Lisa A* and to hold it in position under the cranes within a tolerance of 500 mm for the duration of the lift, by controlling the power and direction of the four thrusters (Fig. 14).

The lifting frame was lowered over the caisson and engaged into the lifting recesses at the base of the walls and the caisson was lifted clear of the barge which then returned to the berth (Figs 15 and 16).

At low tide, the caisson was lowered to the river bed and positioned within a tolerance of 75 mm using electronic distance measuring (EDM) equipment located on survey pillars on the bow of the *Lisa A,* sighting onto targets fixed to the top of the caisson. The data was fed by way of a cable to the on-board computer which gave a direct read-out of the necessary corrections in position to the lift supervisor. Hydraulic jacks which projected through the bottom of the lifting frame were inflated to take load off the crane, to generate sufficient friction to prevent the caisson being displaced by stream-flow or wave forces prior to being properly connected to the river bed.

The two caissons for the main bridge pylons were considerably larger than the other caissons and insufficient friction could be generated by the jacks to resist the lateral loads. Additional temporary restraint was provided by using hydraulic jacks reacting against steel beams cast into holes formed in the rock around the outside of the caisson.

The original method used to found the caisson was to fill a series of independent grout bags suspended under the caisson walls with neat cement grout. Once the grout had reached a strength of 10 MPa it was capable of supporting the weight of the caisson and withstanding the lateral loads. The lifting frames were then released and the *Lisa A* moved to the next location at a suitable high tide.

On some of the later, shallower caissons, hardwood folding wedges were used in lieu of the grout bags with a considerable saving in time and cost.

Expanded metal formwork was used to close the nominal 300 mm gap between the bed and the caisson and the mass concrete plug was poured, preferably in the dry, but using tremie tubes if the circumstances required.

The grout bag technique was modified for those six caissons on the Welsh shore where the bed was never exposed, by making a series of sausage bags which then formed a continuous grout curtain around the perimeter of the caisson. This obviated the need to place formwork underwater to seal the gap and restricted the inflow of sediment under the caisson which had to be removed prior to placing the plug.

Once the mass concrete plug had been completed, the caisson was perfectly stable and the flood valves were closed to allow the placing of the prefabricated reinforcement cages for the upper section and the starter units for the piers. The reinforced section was then constructed in a predetermined sequence of pours, designed to restrict cracking of the shell due to concrete pressures and thermal effects. The temperature of the concrete was monitored using cast-in thermocouples.

Concrete production

Apart from their function as part of the foundations, the caisson shells served to act as permanent formwork for the in situ concrete which was then placed inside the shell. The average caisson required 4500 m^3 while the largest took 6600 m^3. In order to produce the concrete, it was decided to construct a pair of jack-up barges. An existing ten-legged jack-up barge, the *Karlissa*, was purchased and split into two parts. The first section with six legs, the *Karlissa B*, was modified to provide a platform for two Elba EMC105 batching plants with storage silos for aggregates, cementitious materials and water. This barge was moved around the site to each caisson location by the *Sar 3*. Two concrete pumps and placing booms were mounted on the vessel for placing the concrete.

The other section of the barge, the *Karlissa A*, was extended to make another six-legged jack-up barge and four thruster propulsion units were added to enable the platform to navigate the site under its own power. Silos for aggregate, cementitious materials and water were constructed on the barge, as well as conveyors and pumping equipment for handling the materials. The *Karlissa A* was used to ferry all concreting materials from the loading jetty out to the *Karlissa B* and to the batchers on the caissons for the pylons of the cable-stayed bridge. It would berth against the jetty and be loaded during low tide. At high tide it would move across the site and moor against the *Karlissa B*. After discharging the load, it returned to the jetty on the next high tide.

Other marine operations

Two Flexiflote jack-ups were commissioned to provide platforms for cranes, which were used for handling all other materials, such as formwork and reinforcement. These materials were delivered either by road transport along the causeway at low tide or by the three purpose-built workboats.

The transportation of staff and operatives across the site was achieved by using an assortment of floating craft, varying from inflatables to purpose-made 36-man passenger boats.

Acknowledgements

The authors acknowledge with thanks the assistance received from many of their colleagues in the preparation of this paper and in particular Dr B. Jones and the geotechnical team of Halcrow-SEEE at Cardiff, D. H. Mizon of Halcrow-SEEE and C. M. Tong of Halcrow-SEEE. Photographs are reproduced with kind permission of Photographic Engineering Services.

References

1. CHAMBERS S., MADDISON J. D., JONES D. B. and THOMAS A. Aspects of investigations in Trias strata for the second Severn crossing. *11th European Conf. on Soil Mechanics and Foundation Engineering, Copenhagen*, 1995.
2. MADDISON J. D., CHAMBERS S., THOMAS A. and JONES D. B. The determination of shear strength characteristics of Trias and Carboniferous strata from in situ and laboratory testing for the second Severn crossing. *Conf. on Advances in Site Investigation Practice, Institution of Civil Engineers, London*, 1995.
3. HEAD P. R., ILEY P. and BRADLEY E. T. Second Severn crossing—initial studies. *Proc. Instn Civ. Engrs, Civ. Engng, Second Severn Crossing*, 1997, 3–12.
4. ABAQUS. Finite element analysis. Hibbitt, Karlsson and Sorensen, Inc., Pawtucket USA.
5. LARSA. Integrated linear and non-linear software for structural and earthquake engineering. LARSA, Inc., New York.
6. MPILE. Piled foundation analysis and design software. Mott MacDonald, Croydon.

Second Severn crossing— viaduct superstructure and piers

D. H. Mizon, CEng, MICE, MIStructE, MIHT, MIL, *and J. N. Kitchener,* BSc, CEng, FICE

This paper describes the design and construction of the 4·2 km long viaduct for the second Severn motorway bridge between England and Wales. The design of the precast segmental deck was heavily influenced by the Government's requirement of external post-tensioning tendons with free lengths limited to 40% of span. Over 2200 deck units up to 200 t in weight were match-cast on site and erected in balanced cantilever using a 234 m long launching gantry. The paper also describes the precast concrete piers, elastomeric bearings, movement joints, deck finishes and furniture, mechanical and electrical works (including a monorail deck access train and three service gantries) and the toll plaza.

Viaduct deck design

A key design feature of the precast segmental viaduct deck for the second Severn motorway bridge between England and Wales was the use of external post-tensioning tendons, a Government durability requirement. This in turn led to the tri-planar soffit form which suited the layout of the tendons within the box girders and was cheaper to construct than the curved soffit originally proposed (Fig. 1). Additionally, a further innovation resulted from the Government's requirement for reliability at ultimate limit state, which limited the free length of any post-tensioning tendon to 40% of the span. Although the tender design showed numerous additional anchorages to limit tendon length to comply with this requirement, the final design incorporated intermediate blocking devices (IBDs). By preventing longitudinal movement of a tendon relative to the deck box girder, these performed a similar task to the additional anchorages but reduced the number of holes through the

Fig. 1. General view of the viaduct

David Mizon, a design manager at Halcrow-SEEE

Neil Kitchener, chief engineer at Laing-GTM

diaphragms and the number of tendons.

The three features significantly affected the development of the design, precasting system and construction methods. External tendons simplified deck unit production and enabled thinner webs and flanges to be used but necessitated a deeper box girder owing to the reduced tendon eccentricity. The section depths of 7·0 m at the piers, reducing to 3·95 m for the central section, were somewhat deeper than those shown in the tender design and much effort was spent optimizing these since the central constant-depth section had such an impact on the appearance of the viaduct.

A twin box-girder deck was preferred to single box sections, mainly to reduce segment weights and to remove the need for long cantilevers with transverse post-tensioning. The deck slab was made transversely continuous by a 2 m wide in situ stitch slab between the two 15·6 m wide box girders. The stitch slabs had to be checked for the effects of differential movements of the two box girders due to live-load deflections and thermal effects.

It was a Government requirement that no openings should be made through the deck for maintenance access, this being provided from below by an access train and gantries which run on rails under the central reservation in the void between the two boxes.

Longitudinal deck movement joints were located mid-span at approximately 500 m centres. The mid-span location suited balanced cantilever construction and enabled the span module to be maintained, except at each abutment.

The longitudinal joint spacing adopted is convenient for medium-range movement joints but was also limited by a Government requirement as part of the overall philosophy to limit progressive collapse. Structural discontinuity at the mid-span joints and the need for a concentration of tendon anchorages required robust diaphragms at the end of each cantilever. They also house 'mortise and tenon' type units needed to transmit vertical and horizontal shear forces from one half-span cantilever to the other. These units permit rotation and longitudinal relative deck movements by the provision of elastomeric bearings with stainless steel–polytetrafluoroethylene (PTFE) sliding surfaces. The post-tensioning in the cantilevers each side of a joint was arranged to prevent creep deflections downwards which would lead to a 'gull's wing' effect on the vertical alignment.

As a consequence of these special characteristics, the end half-span cantilever deck units were non-standard. The tenon units were constructed by casting the rectangular tenons as a second-stage operation. The units were subsequently slid into position and the tenon located between the bearings in the rectangular mortise openings through the diaphragms in the deck segment at the opposite side of the joint. At the junctions with the cable-stayed bridge, steel beams post-tensioned to the end diaphragms were located by bearings fixed in fabricated steelwork frames at the ends of the backspan deck steelwork. These transmit vertical dead-load reactions from the cable-stayed bridge in addition to live-load and other effects. The dead-load reactions were controlled by constructing the cable-stayed bridge deck to a predetermined profile lower than the viaduct and introducing the load into the viaduct deck by incremental jacking. The cable-stayed bridge deck profile was determined by selecting the backspan cable-stay forces to produce the required effects.

Two elastomeric bearings beneath each box girder transmit vertical and horizontal loads from the deck to the piers. Thus the deck floats with no fixed point, enabling thermal length changes and longitudinal restraint, braking and seismic forces to be distributed over several piers.

The longitudinal prestress was standardized with tendons consisting of 19 x 15·7 mm 1860 MPa (low relaxation) strands.

The tendons are classified into four groups.

- *Cantilever tendons*—installed during erection of the units and anchored in anchorage blocks

Fig. 2 (below). Internal view of girder box viaduct showing tendon layout

Fig. 3 (bottom). Viaduct tendon layout

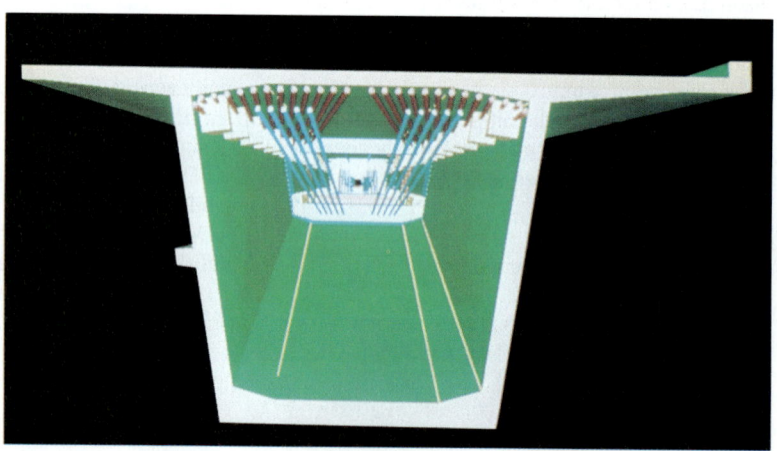

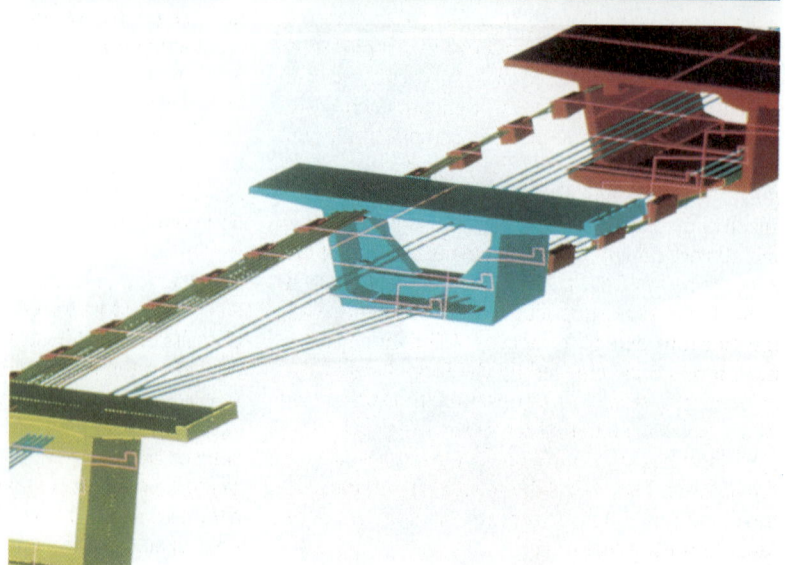

VIADUCT SUPERSTRUCTURE AND PIERS

in the top corner of the deck segments. Each segment has a pair of cantilever tendons except for the last one (segment 13). The cantilever tendons of segment 12, needed during construction, were removed after each mid-span in situ joint was cast and the continuity tendons stressed.

- *Mid-span lower tendons*—located above the bottom flange of the central region of each span. These are anchored in anchorage blocks in the bottom corner of each segment.
- *Mid-span upper tendons*—located just below the top flange soffit in the mid-span region between the deviator beams of segment 8.
- *Continuity tendons*—extending over two or three spans and anchored in the diaphragm of a pier or joint segment. To provide the desired variation of eccentricities along the span, direction changes for the tendons are achieved by deviator tubes in the segment-on-pier (SOP) and deviator segments which are located where the deck soffit changes direction. Figs 2 and 3 show a typical tendon layout.

The intermediate blocking devices mentioned above were located to comply with the 40% span requirement and consist of purpose-made steel assemblies bolted to the segments with anchorage-type wedges to grip the tendon and hold it in position longitudinally. These assemblies were designed for a maximum working load of approximately 400 kN, representing the maximum force needed to ensure strain compatibility between the tendon and segment at ultimate load. The design and performance of the IBDs were proved by a programme of fatigue and capacity testing. Details of the method of analysis used for the ultimate limit state and the determination of the design loads for the IBDs are given by Fletcher *et al.*[1] This is based on a non-linear elasto-plastic method and provides a basis for calculating the load factor at collapse for externally prestressed multi-span decks.

To comply with the segment weight limitation of 200 t, the special segments over each pier were designed with in situ sections which were placed once the unit was in position. Beneath each SOP, a temporary arrangement of jacks and reinforced concrete stools supported the deck during cantilever construction. These were located in the space above each pier and were designed to transmit all vertical and horizontal loads and out-of-balance moments from the deck to the pier. No additional temporary support frames were needed at the pier tops except for access. Adjustment to deck levels was achieved by the jacks located between the stools. The construction loading case was a major consideration for the design of the pier. The high concentrations of load caused by the jacks and stools on the deck unit and pier head resulted in high concentrations of reinforcement in these areas.

At the Avon end of the crossing, the viaducts widen by up to 13 m over the last 240 m to accommodate the on and off slip roads to the M49 motorway. In this region, the deck segments diverge and the deck between them consists of an in situ reinforced concrete slab supported on 2 m deep precast reinforced concrete beams which span between the inner webs of the segments. In order to accommodate the maintenance train which operates between the segments, a large rectangular opening was formed in the central region of each precast beam.

Three-dimensional CAD drawings were produced during design development showing the layout of the tendons and anchorage blocks in the box girder deck. These were useful for investigating various tendon layouts and planning construction and access arrangements for future maintenance in advance of construction.

The length of the viaduct is 2076·9 m on the west (Gwent) side and 2101·5 m on the east (Avon) side. The span layouts are

- west viaduct: 65·4 m + (20 × 98·12 m) + 49·06 m (half-span cantilever)
- east viaduct: 32 m + 58 m + (20 × 98·12 m) + 49·06 m (half-span cantilever).

At the time of construction the viaducts were the longest constructed to date using total external prestressing.

The chosen construction method required that each time a deck unit was placed, it was brought along the already constructed deck on a multi-bogey transporter. The completed deck is designed to withstand HA and 45 units of HB loading but the longitudinal and transverse design had to be checked for the effects of the unit transporter load. In addition, the effects of the overhead gantry and the various locations for the supports on the deck were further load cases which had to be checked.

During balanced cantilevering operations the deck segment over each pier was vertically post-tensioned onto the pier. Both ends of the tendons were anchored on the deck and passed through U-tubes cast into the pier. This system, in conjunction with the temporary jacks and stools on the pier top, provided the required stability against overturning for out-of-balance deck erection conditions.

Viaduct deck construction

A standard balanced cantilever span comprised 27 units of equal length, the centre segment supported on the pier and 13 units on either side. The first eight units varied in height and the remaining units were of constant height. The chosen construction method involved a purpose-made erection gantry which operated above the deck and launched itself forward to construct each span, with the units being delivered by transporters which ran along the completed deck from the shore. The launching gantry was also able to traverse sideways, enabling both boxes to be erected simultaneously (Figs 4 and 5).

Unit manufacture

Over 1100 units were manufactured on each side of the river in purpose-made under-cover pre-casting facilities containing a 205 t gantry crane which was used to move the units between the five casting bays.

The units were match-cast against their immediate neighbours in a series of five moulds. Mould 1 was dedicated to the construction of the SOP unit. Moulds 2 and 3 were designed to be adjustable and were used for the manufacture of the seven variable-height units. Mould 4 was used for the transition unit 8 and mould 5 was used for the constant-height units 9 to 13.

Each casting bay consisted of a set of hydraulically operated steel moulds in which a unit was cast and a platform which allowed the relative

Fig. 4. Gwent viaduct showing balanced cantilever construction

Fig. 5. Support beam allowed transverse movement of launching gantry

alignment of the previous unit to be adjusted. The correct alignment was achieved by accurately surveying the relative geometry of four control stations on the top of each units. The results were fed into a sophisticated computer program which contained the required construction geometry and compensated for casting errors in the previous units. It calculated the necessary adjustments in orientation so that the new unit was cast in the correct relative orientation to achieve the desired profile after erection.

The trapezoidal units were generally cast in one pour; however, certain units such as the SOP and unit 8 were constructed as a shell and the complicated internal diaphragms were added as a second-stage operation once the unit had been removed from the casting shed and was off the critical path.

The nominal strength requirement for the concrete was only 60 MPa but in order to achieve the required early strength for striking formwork and lifting the unit, an average concrete strength of 80 MPa was achieved using 480 kg of cementitious material with 28% of blast furnace slag.

The early strength of the concrete in each unit was monitored by maturity measurements, calibrated monthly, to ensure that the formwork was not struck or the units lifted too early. Target strengths of 12 MPa for striking formwork and 17·5 MPa for lifting units were regularly achieved within 12 h and 18 h respectively.

In order to optimize productivity on the line, the reinforcement was prefabricated wherever possible (Fig. 6). Hot water was used in the production of concrete during the winter months and limited heating was applied, particularly during cold spells. It proved possible to manufacture up to 20 units from the five moulds in a six-day week in all but the worst weather conditions.

The construction of the units in a single pour and the careful monitoring of concrete strength which allowed early striking of the formwork prevented differential shrinkage and constraint during curing. This contributed to an almost total absence of cracks in the units.

Unit transport

Prior to erection, the mating surfaces of the units were prepared using hot, high-pressure water and the lifting eyes were installed. The formwork and reinforcement for the third-stage concreting operation on the SOP were also fixed.

Fig. 6. Prefabrication of viaduct deck unit reinforcement

Fig. 7. Viaduct launching gantry

Fig. 8. Lifting viaduct segment-on-pier (SOP)

The straddle carrier picked up the unit from the storage yard and delivered it to a purpose-made loading bay from where it was picked up by the self-loading multi-wheeled Nicholas trailers which transported the unit along the completed deck from the yard to the rear of the launching gantry.

Unit erection

All the deck units were erected using the launching gantry, which was designed as a self-contained machine for placing segments on both carriageways. The size of the gantry was minimized by using deck units as counterweights where necessary during the launching operations and restricting the erection weight of the SOP unit to 200 t, although this meant adding a further 180 t of concrete once it had been erected.

The gantry consisted of a pair of 234 m long trusses, two primary transverse supports, secondary supports on the front and rear of the truss and a pair of crab units. When placing units, the rear support was fixed above the last SOP on the completed deck and the front support was stressed through the next SOP onto the top of the pier to ensure stability (Fig. 7).

On completion of a span, the truss was launched forward using the capstan winch on the front crab until the front leg could be placed on a temporary bracket which cantilevered forward

VIADUCT SUPERSTRUCTURE AND PIERS

Fig. 9. Third stage concreting of viaduct segment-on-pier

Fig. 10. Gluing the viaduct deck unit

from the front face of the pier, and the SOP was then placed in position on its temporary supports (Fig. 8). The truss was winched backwards and then moved laterally to enable it to place the SOP for the other carriageway. Once both units were in position and correctly aligned, the temporary concrete bearings were grouted into position and the third-stage concreting operation on the SOP units was then carried out (Fig. 9). Both primary support girders were moved progressively forward until the rear support was sitting above the last pier of the completed span and the front support was above the new SOP. The base of each support girder was then stressed through the SOP onto the pier to ensure stability.

The standard erection cycle then started. As each unit was delivered and lowered into position by the crab, epoxy glue was applied by gloved hand to the joint surfaces (Fig. 10). It was aligned and temporarily prestressed to the previous unit using six 50 mm Macalloy bars which were placed in temporary ducts cast into the flanges (Fig. 11). Once a balanced pair of units had been erected, the permanent cantilever tendons at the top of the segments were stressed and the temporary stressing was released, allowing the Macalloy bars to be used for the erection of the next unit.

When all 27 units had been erected on both carriageways, the temporary stressing through the front supports was detensioned and the hydraulic jacks under the SOP were operated to adjust the alignment of the balanced cantilever span. The permanent elastomeric bearings were grouted into position and the vertical load transferred from the temporary jacks to the permanent elastomeric bearings. The nominal 200 mm gap between the old and the new span was concreted and the permanent continuity tendons above the bottom flange were then stressed to enable the gantry to be launched forward to start the next span. The second-stage continuity stressing was installed later in the erection cycle (Fig. 12).

The alignment of the completed deck is very good and the fact that shimming was only used to correct the geometry on four occasions is a tribute to the attention paid to the quality of the survey work in the precast yard and the accuracy of the setting-out data supplied by the designer for the construction phases.

Incident

On 2 June 1994, the SOP unit for pier N6 was being erected by the Gwent gantry. The unit was lifted by the crab in the normal manner and conveyed forward. When it reached the rear support, the operator realized that it had not been lifted high enough to clear the support frame and the forward motion was stopped to allow the unit to be raised. The slope of the gantry was approximately 2·5% and the crab started to roll backwards. The operator could not stop the backward movement of the unit, owing to a malfunction in the braking system. It eventually hit the stops on the end of the truss, which collapsed, dropping the unit 5 m onto the completed deck. It broke through the top slab and came to rest on the bottom slab, which was also severely damaged. Fortunately, nobody was injured during the incident (Fig. 13).

Nevertheless, the majority of the span remained intact. Following a comprehensive physical and analytical examination of the damage by the designer, including a check on the dynamic loading effects of the falling segment on the whole structure, the unit was removed from the deck after the structure had been made safe. The damaged span was repaired to the satisfaction of the designer, the checker and the Government agent. The gantry was closely inspected over its full length and extensive repair work was carried out. The braking system on the electrically powered rear crab was replaced with upgraded units. A replacement unit was manufactured in the mould and erected by the repaired gantry some three months later. Although this unit was not match-cast, the erected geometry was acceptable.

It was fortuitous that the impact occurred in the mid-span region where the effects were less severe than they would have been in other regions, owing to the prestress layout and the construction sequence which resulted in two cantilever ends joining at this point. In addition, external prestress made repair work simple. For

Fig. 11 (left). Temporary stressing of the deck units

Fig. 12 (above). Internal view of the viaduct showing the stressing ducts and deviator unit

Fig. 13. Pier unit embedded in the deck following the Gwent incident

VIADUCT SUPERSTRUCTURE AND PIERS

further details see the *New Civil Engineer* second Severn crossing supplement.[2]

Stressing

The 110 mm high-density polyethylene (HDPE) ducts for the external unbonded stressing were temporarily supported on frames suspended from the soffit of each unit. The ducts were made continuous between the anchorages by using electro-fusion couplers and passed through the holes in the deviator blocks formed with the pre-bent galvanized steel ducting. O-ring seals were used to connect the duct to the anchorages and to the body of the IBDs.

Strands were placed individually by using a strand pusher mounted on the deck, feeding through holes in the top slab. All tendons, even the 250 m long continuity tendons, were stressed from one end using a multi-strand jack and the IBDs were then clamped around the tendon and bolted onto the segment. Within four weeks of completing the stressing, the ducts were injected with hot wax to complete the protection system.

End span construction

The configurations of the end spans at both abutments were non-standard and could not be erected in balanced cantilever. Military trestling was used to support a pair of beams and the extra units were placed on skates using the launching gantry operating from behind the abutment (Fig. 14). When all the units were in position, they were glued and temporarily stressed together before being connected to the standard balanced cantilever span which had been erected on the first pier.

The first four spans of the Avon viaduct were further complicated by the divergence of the boxes to accommodate the slip roads and the sequence of erection for the units in the area was also non-standard. The box girder for the south carriageway was erected by the gantry as far as pier N45. The gantry then travelled back to the abutment and traversed northwards before erecting the north box up to N45. The support frames were then modified so that the gantry could thereafter construct both boxes simultaneously.

The majority of this length was outside the sea wall and falsework for the central infill slab and the precast beams could not be supported from below. The central sections of the reinforced concrete beams used to carry the slab were precast in the construction yard and lifted from a barge by a crane operating on the deck. They were held in position by a frame spanning between the boxes while the in situ stitch connection was made to both webs (Fig. 15). The runway beams for the permanent gantry were fixed to the underside of the beams and used to support a temporary access platform for erecting and striking the formwork for the 250 mm thick slab, which was also supported from above by a frame spanning between the two boxes.

Movement joint construction

At each movement joint, the sequence of erection differed from the standard mid-span connection. The last unit (unit 13) placed on the span before the joint was the special tenon unit. After unit 11 had been erected on the span beyond the joint, the mortise unit 13 was lifted by gantry, placed onto the tenon on temporary bearings and temporarily stressed back to the tenon unit before being released by the gantry. After unit 12 had been erected and stressed, the gantry picked up the mortise unit and repositioned it on the end of the new span.

After the span had been aligned, the sliding elastomeric permanent bearings above and below each tenon were installed and preloaded to provide the necessary load transfer between the ends of the cantilevers, yet permit longitudinal movement.

At the junction with the main bridge, the tenon was provided by a steel beam post-tensioned onto the end of the viaduct unit, which was erected in the normal manner without the tenon. The tenon was inserted into the mortise in the last unit of

Fig. 14 (top). Erection of the Avon end-span deck units

Fig. 15 (above). Precast concrete beams at the Avon end span

the main bridge and temporarily fixed in position. This unit was erected in the normal manner by the double shearlegs. The end of the main bridge was lifted up to its final level by a beam-and-jack arrangement attached to the end of the viaduct. The tenon was then slid into its final position and stressed onto the viaduct, and the permanent bearings were installed and fixed into position before the temporary beam support was released.

Viaduct pier design

At tender stage, all the piers for the viaduct were to be constructed as in situ reinforced concrete columns. However, the large number of fairly small pours involved in their construction would have seriously compromised the output of the marine batching plant and threatened the overall programme for the foundations. The work would also have been labour-intensive and subject to the problems of bad weather and working in the hostile marine environment. A precast solution was therefore developed for the 37 piers which could not be constructed by shore-based plant.

The precast piers, varying in height from 7 m to 40 m with an overall plan size of 6 m x 3·5 m with 500 mm thick walls, were vertically post-tensioned onto the caissons. In order to standardize manufacture, each pier included a 1 m deep starter unit which was cast into the top of the caisson and a 5·5 m high pier-head unit which supported plinths for the viaduct bearings. In order to make up the required height for each pier, the balance was made up by using standard 6·5 m tall units with one unit of variable height. The maximum weight of the units was frozen at 180 t in order to enable the design of the handling equipment to be finalized.

The prestressing details needed careful development to ensure a system that was both durable and buildable. The solution adopted, using buried unbonded tendons located within the precast units, is considered to possess the desired durability aspects while being the most efficient structurally, because the tendons are located in the zone of maximum tension when the pier is subjected to moments. A trial pier was constructed to check that it was feasible to install, remove and replace the tendons.

A double corrosion-protection system is provided by using HDPE ducts and U-bends made from steel pipes filled with wax. The joints between the precast segments, especially those in splash zones, are very vulnerable to ingress from chlorides. In addition, the Government's requirements stipulated that prestressing tendons used in segmental construction should be replaceable. The ducts are made continuous and waterproof by using electrofusion couplers and encasing the joints in the ducts with in situ concrete.

High compressive stresses in the piers result from combinations of vertical axial load, vertical prestress and applied moments from wind, bearing forces, seismic loading, etc. Concrete of 60 MPa grade was used for the precast segments and for the encasing concrete of the in situ duct joint.

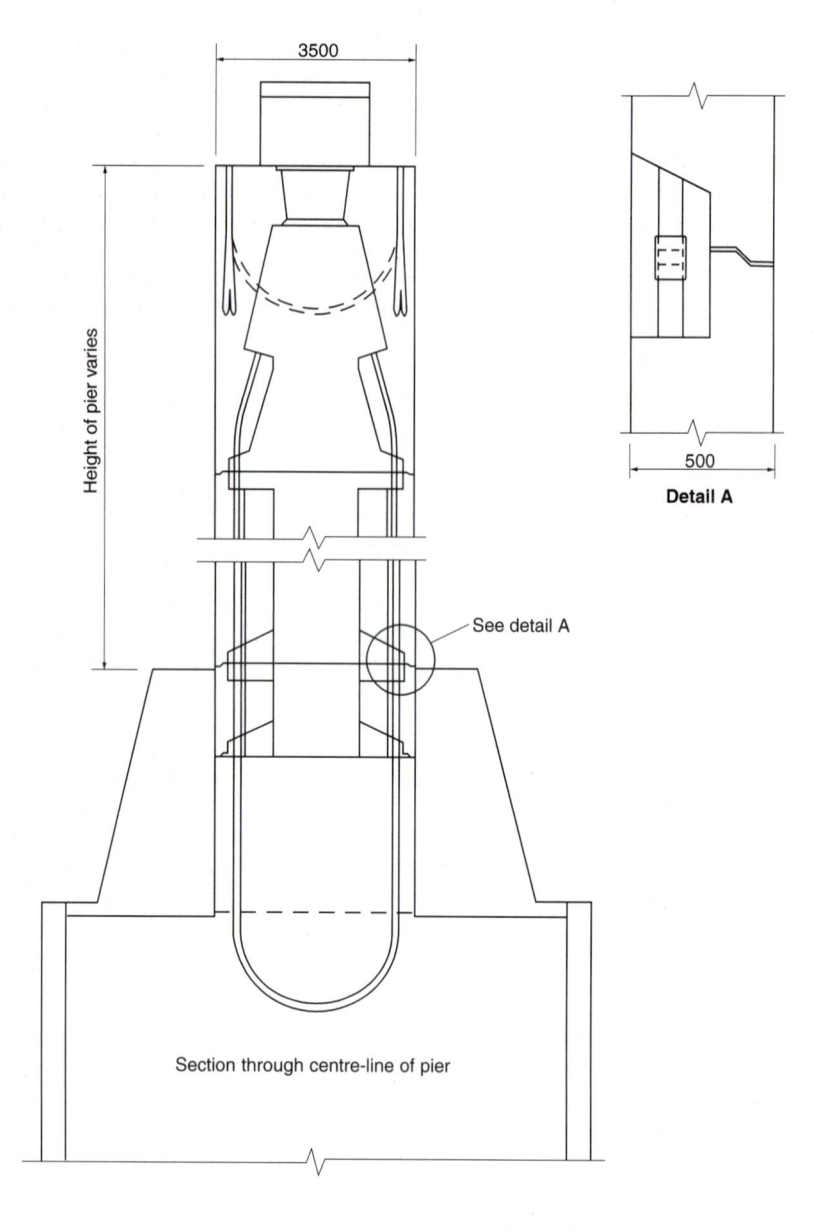

The prestress anchorages are located inside the top unit of each pier and can be conveniently accessed using steel mesh platforms. In the upper reinforced concrete region of the caisson infill the tendons are located in steel U-bends which connect onto the vertical HDPE ducts. See Fig. 16 for a typical section through the pier.

Pier stability against overturning from wind was checked for critical construction stages when the pier was free-standing without prestress. No temporary prestress or additional temporary works were needed for the piers, the tallest of which was 45 m above sea-bed level. The design wind speed for this temporary condition was limited to that appropriate to a return period of 10 years.

The prestressing of many of the shorter piers was governed by construction loads. The balanced cantilever viaduct deck erection method

Fig. 16. Section through the precast pier (all dimensions in mm)

VIADUCT SUPERSTRUCTURE AND PIERS

imposed high moments on the piers. The holding-down arrangements at the piers to counteract out-of-balance cantilever moments from the deck required vertical temporary prestress from the deck unit top into the upper region of the pier, together with temporary stools and jacks supported on the pier top. The width of the pier head at 3·5 m minimum and the high stress levels generated by the cantilever erection necessitated close examination of the stresses in the region using finite element modelling. The resulting high densities of reinforcement necessitated great care when fixing to ensure good-quality concrete placing using 10 mm aggregate, and correct positioning of the reinforcing bars.

Bearings

The viaduct deck was designed with movement joints in the middle of every fifth span. Elastomeric bearings were chosen in order to minimize the horizontal load on any one pier due to a quasi-static seismic loading equivalent to 0·06g, except at the abutments and movement joints in-span, where stainless steel/PTFE sliding surfaces were incorporated in order to accommodate longitudinal deck movements and also limit seismic loads. The stiffnesses of the bearings in each group of five were varied to equalize the lateral load distribution on the substructure as far as possible.

The outermost piers of each group were subject to higher thermal and time-dependent induced movements such as creep and shrinkage and less stiff bearings were used in these locations to reduce the amount of stress to the system. The overall stiffness of the system was calculated taking into account the stiffness of the bearing together with the piers and foundations. The stiffness of the bearings was controlled primarily through adjusting the overall depth of elastomer, which is inversely proportional to the shear stiffness.

The seismic design loading led to very large bearings of 1·3 m x 1·3 m plan area with depths of up to 500 mm. Lateral deformations were estimated to be between 150 mm and 200 mm for the seismic load case, while under normal loading conditions, the maximum deformation was expected to be in the region of 80 mm. A fail-safe shear-key device was incorporated into the pier head–deck detail to prevent excessive deflections in the case of a seismic event.

A disadvantage of using the bearing stiffness as a way of controlling lateral load distribution was that it became a critical factor requiring a fairly tight tolerance. BS 5400 Part 9 normally specifies elastomeric bearing stiffness to 20%. Working to this level of tolerance would have nullified the whole design approach and so new tolerance criteria were established which effectively reduced the tolerance to ±7·5% of the design value.

The requirements of the design specification were met by a carefully planned manufacturing programme, the key elements of which were the production and testing of prototype bearings and tight control over the sourcing of the batches of natural rubber, most of which came from Indonesia. Shear tests were carried out on every bearing for normal in-service loads. In addition, the prototype bearings were subjected to a full series of tests including a shear load representing full seismic loads. Of the 180 bearings manufactured, only one failed to meet the requirements of the specification and was replaced.

Viaduct pier construction

The pier units were precast in the Avon precast shed in purpose-made steel moulds using vertical match-casting techniques. The HDPE ducts were cast into the walls for the unbonded post-tensioning. The top and bottom of each unit were profiled such that there was a 200 mm wide contact surface and a 300 mm wide x 400 mm deep zone which was filled with reinforcement and in situ concrete once the units had been erected and the stressing ducts connected. All reinforcement in this in situ joint was epoxy-coated to improve durability.

The steel U-tube anchorages, the insides of which had been treated with an anti-corrosion compound during manufacture, were incorporated into the reinforcement cage for the last caisson pour. The precast pier starter units were suspended from a purpose-made jig and the ducts connected to the anchorage tubes using electrofusion couplers. Great precision in setting each unit was essential in order to ensure that the pier could be constructed

Fig. 17. Precast pier unit being loaded for transportation from the storage yard

within tolerance. The ends of the ducts and the injection pipes were kept sealed at all times to prevent ingress of sea water and construction debris.

Prior to transportation out to the river, the mating surfaces of each unit were prepared, the electrofusion couplers were slipped over the top of each duct and the internal ladders and platforms were fitted. Up to six units were loaded by the straddle carrier (Fig. 17) onto a sliding frame fixed onto the top of the Lampson crawlers which then tracked onto the *Sar 3* self-propelled barge at low tide.

At high tide, the *Sar 3* sailed out to the *Jay Robertson*, a six-legged jack-up barge on which was mounted a 200 t Manitowoc 4100 ringer crane, which it had earlier positioned alongside the caisson. The *Sar 3* maintained station next to the *Jay Robertson* using its thrusters while the units were lifted off and stored temporarily on the deck to allow the *Sar 3* to return to the jetty before low tide (Fig. 18).

Epoxy glue was applied by gloved hand to the upper mating surfaces of the starter unit and the first unit was placed in position and checked for accuracy. The process was repeated, using shims and additional thickness of epoxy glue if necessary to maintain the alignment, until the pier head unit had been placed. The weight of each unit was sufficient to compact the glue adequately during curing and to ensure that the partially completed pier was stable under extreme wind conditions (Fig. 19).

The HDPE ducts were welded together by the electrofusion couplers, the in situ joints between the units were reinforced and concreted using letterbox shutters and the temporary internal working platforms were removed.

A purpose-made stressing platform spanning between the pair of piers was lifted into position. Each tendon, made up on the platform from 19 x 15·7 mm 1860 MPa strands, was winched into the ducts, after they had been thoroughly cleaned to remove any water or debris at the bottom of the U-tube. The tendons were stressed from both ends simultaneously and after the end caps had been fitted hot Denso wax was pumped up from the bottom of each duct to provide the corrosion protection.

The early bearing plinths were cast in situ but a later improvement was the development of a precast unit which was lifted into position and grouted onto dowels projecting from the top of the pier head unit.

On completion of the piers, the profiled chapeau, designed to protect the base of the piers from ship impact, was cast onto the top of the caisson. Additional protection against ship impact was provided by filling the lower parts of the more vulnerable piers with mass concrete once the viaduct deck had been erected and access was more readily available.

Fig. 18. Precast pier and viaduct deck erection

VIADUCT SUPERSTRUCTURE AND PIERS

Bridge deck finishes and furniture

Apart from the widened section to accommodate slip roads at the east end, the crossing comprises a dual three-lane carriageway with a hard shoulder and emergency walkways on each side.

The central reserve has a single-rail box-beam safety fence while double-rail box-beam safety fences separate the hard shoulder from the emergency walkway. The 3 m high windshielding is located on the outer edges of the walkway and consists of aluminium posts with three profiled rails spaced vertically to give 50% porosity. Parapet rails attached to the inside of the windshielding posts give additional protection to P1 standard against errant vehicles passing over the edges of the bridge.

The windshielding posts are anchored to the concrete deck at the cross-girder locations on the main bridge and on ribs on the viaduct deck units which provide transverse stiffness to resist the high forces from wind and parapet impact.

The roadway surfacing consists of 100 mm of hot-rolled asphalt placed in two layers onto a 20 mm sand–asphalt carpet protecting the polyurethane-sprayed deck waterproof membrane. The subsurface drainage consists of aluminium inverted U-tubes located along the edges and discharging into the deck surface-water drainage gullies. These discharge directly into the estuary through PVC pipes, except over land where specially developed fibreglass U-channels take surface water into the road drainage system at each abutment. These are bolted directly under each cantilever soffit and are pigmented to suit the overall bridge colour scheme.

Elastomer-covered steel plate expansion joints capable of accommodating up to 500 mm of movement were installed after the surfacing had been laid.

The roadway lighting columns are located along the edges of the crossing in the emergency walkway. A system of closed-circuit television surveillance units and emergency telephones is provided along the route to assist drivers in difficulty.

Generally, all metallic components of the bridge furniture are coloured to suit the architect's chosen colour scheme.

Mechanical and electrical works

The emphasis on durability and the provision of good maintenance inspection facilities resulted in a number of features involving significant mechanical and electrical items. These had to be considered in the early design stages and incorporated into the design as it developed.

Fig. 19. Precast pier erection

The principal mechanical and electrical items are as follows
- monorail access train and stations suspended beneath the bridge deck
- access gantry for the main span of the cable-stayed bridge
- one access gantry for each viaduct, which also services the backspans of the cable-stayed bridge
- access lifts in each of the four pylon legs
- 11 kV power supply cables within the bridge deck
- a dry main on the bridge deck for fire protection
- a compressed-air main underneath the cable-stayed bridge deck
- extensive access provisions in all parts of the structure, such as ladders, platforms and walkways
- lighting within the bridge deck, pylons, piers, etc.
- navigation warning lights on the pylons and piers
- a foghorn on one of the pylons
- aircraft warning lights on the pylons
- street lighting, communication, dot matrix signs, closed-circuit television, emergency telephones, etc.

The access train operates over the whole length of the crossing between the east and west abutments and is accessed by stations located at each end. Intermediate steel platforms suspended from the bridge deck or on the pylon allow access into the structure.

The train is designed to carry personnel and has a trailer for carrying equipment and materials.

A conventional underslung steel gantry supported on rails beneath the deck and on the deck slab edges provides access below the main span between the pylons. However, the access gantries which service the backspans and viaducts are a novel design consisting of a central rigid carriage from which a rotating steel truss access beam is suspended. The rotational facilities allow the gantry to pass between the piers by turning the access beam through 90°. Hydraulically operated telescopic platforms permit access to the different levels beneath the deck.

The wheels on the central carriage of the gantry run on concrete corbels which were cast with the precast deck box-girder units.

On the widened end of the east viaduct the gantry carriage runs on steel rails suspended from the deck slab and cross-beams. Likewise, steel rails support the carriage over the backspans of the cable-stayed bridge.

Within the pylon legs electrically operated rack and pinion lifts provide access from the caissons to the top of each cable-stay anchorage zone. In addition, a robust system of short steel ladders with platforms provides access over the full height of the pylon.

The power cables are located on steel cable trays within the viaduct box girders and on trays supported from the deck cross-girders on the cable-stayed bridge. To permit cabling to be carried out in the future, permanent steel walkways are provided adjacent to the power cables within the open deck of the cable-stayed bridge.

Toll plaza

The Government's firm requirement stipulated that a single toll-collecting facility should be located on the Gwent side of the crossing at a designated location some 3 km from the west abutment. Single-way tolling proposed by the concessionaire reduced the number of toll booths and operating costs.

Between the site of the toll plaza and the bridge, the motorway was constructed as part of the Gwent approach road contract.

The toll plaza site is located on flat ground with some 9 m depth of peat in places and a high water-table. Various soil replacement and ground improvement techniques were examined, resulting in a solution using 11 000 vibrated concrete columns (VCCs) over the whole area.

All the structures are also supported on VCCs, spaced to suit concentrations of load. A geotextile gravel blanket distributes structure and highway loading onto the VCCs. The spacing of the VCCs was optimized by carrying out load tests on trial sections adjacent to the site. Maddison et al.[3] provide further information on VCCs.

The toll-collecting facility consists of a toll booth, canopy structure and administrative offices, with the necessary roadway widening and associated road and drainage works. The toll booth canopy has a steel-bar suspended space-frame roof and column structure of unusual design.

Acknowledgements

The authors acknowledge with thanks the assistance received from many of their colleagues in the preparation of this paper and in particular Y. Maury and Dr H. Khadivi of Halcrow-SEEE, Paris, and M. S. Fletcher and C. M. Tong of Halcrow-SEEE, UK. Photographs are reproduced with kind permission of Photographic Engineering Services.

References

1. FLETCHER M. S., MAURY Y. and KHADIVI H. Second Severn crossing approach viaducts: an example of total external prestressing. *FIP Symp. on Post-tensioned Concrete Structures*, 1996.
2. Second Severn crossing supplement. *New Civil Engineer*, May, 1996.
3. MADDISON J. D., JONES D. B, BELL A. L. and JENNER C. G. Design and performance of an embankment supported using low strength geogrids and vibro concrete columns. *Proc. 1st European Geosynthetics Conf., EuroGeo 1, Geosynthetics: Applications, Design and Construction*, Maastricht, 1996.

Second Severn crossing—cable-stayed bridge

D. H. Mizon, CEng, MICE, MIStructE, MIHT, MIL, *N. Smith*, BSc, CEng, MICE, and *A. J. Yeoward*, BSc, MSc, DIC, CEng, MICE

This paper describes the design and construction of the 456 m cable-stayed span of the second Severn motorway bridge between England and Wales. Including backspans, the cable-stayed section of the crossing is over 900 m long. The 35 m wide deck is made from steel plate girders and a composite reinforced concrete slab, prefabricated on shore and erected using balanced cantilever methods. The deck is supported by 240 cables from two twin-leg, reinforced and pre-stressed concrete pylons 149 m high. Cable vibrations experienced during construction were eliminated by the installation of secondary cables.

CABLE-STAYED BRIDGE

Proc. Instn Civ. Engrs, Civ. Engng, Second Severn Crossing, 1997, 49-63

Paper 11443

David Mizon, a design manager at Halcrow-SEEE

Neil Smith, a senior engineer at Laing-GTM

Andrew Yeoward, cable-stayed bridge design team leader at Halcrow-SEEE

General description

The 900 m long cable-stayed structure which crosses the channel known as The Shoots in the middle of the Severn estuary is the focal point of the 5 km second motorway crossing between England and Wales.

The cable-stayed bridge is symmetrical, with two concrete portal frame pylons with vertical legs from which radiate four planes of 60 cables in a modified fan formation, making a total of 240 cables.

The 34·6 m wide deck is located some 40 m above high water level and consists of steel plate girders with a composite reinforced concrete deck slab. The cables pass through openings in the deck and are anchored on the outside of the deck steelwork.

The main span of 456 m was needed to cross The Shoots channel while maintaining sufficient distance from the foundations to the steep side slopes of the channel. Additional stiffness to the structure is provided by two sets of backspan piers on each side located at 98·12 m and 196·24 m from each pylon (Fig. 1).

The articulation was developed following investigations into various ways of fixing the bridge deck against longitudinal movement. By locating hydraulic shock transmission units at each pylon, longitudinal movements due to temperature changes and creep can take place with low restraint forces but instantaneous longitudinal forces from braking, live load changes or seismic action are transferred from the deck equally to both sets of pylons. The units also incorporate elastomeric springs, designed to ensure that the deck remains sensibly central about the pylons.

Vertical forces from the deck are transferred by sliding pot bearings located on the lower crossbeam and elastomeric bearings transfer lateral wind forces to the pylons at the sides of the deck. At each backspan pier, longitudinal deck movements are accommodated by guided sliding pot bearings which transmit vertical loads and wind loads to the backspan piers. The deck is tied

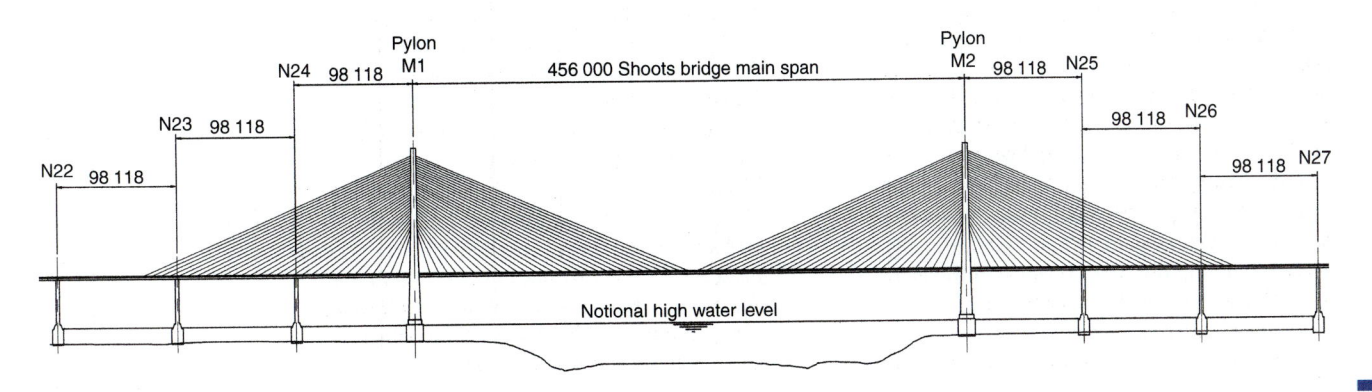

Fig. 1. General arrangement of the cable-stayed bridge (all dimensions in mm)

down at each backspan pier by four vertical tendons anchored in the lower region of each pier.

At each end of the cable-stayed bridge, the vertical and lateral shear forces are transferred to the viaduct cantilevers by pot bearings and sliding elastomeric bearings located on steel beams post-tensioned back to the viaduct deck and located in rectangular openings in cross-members under the deck.

Design of pylons and backspan piers

The twin-leg concrete pylons each have two cross-beams for transverse stiffness. The pylon legs are hollow, with lifts and steel ladders providing internal access from caisson level to an entry at deck level and the top of the pylon. The upper regions of the legs are post-tensioned vertically. Generally, the pylon walls are of 60 MPa reinforced concrete but in the upper region, 65 MPa concrete was used. Tensile splitting forces between the cable anchorages on opposite sides of the pylon are resisted by precast tie beams fixed between the walls, post-tensioned by Macalloy bars.

The pylon tops are 149 m above river-bed level and 101 m above deck level, the latter height being 22% of the main span length (Fig. 2). The cross-beams are hollow, precast reinforced concrete box girders. Longitudinal post-tensioning was used to reduce the reinforcement in the upper cross-beams. These are the only elements of the project with buried, grouted tendons and to ensure appropriate durability, the recommendations of the Concrete Society[1] were followed for the design, construction and grouting of these tendons.

Transverse to the bridge, the pylon leg widths are uniform at 4·0 m but in the longitudinal direction the base width is 10·2 m reducing to 5·4 m. Pylon proportions were the subject of much scrutiny by the project architect and designer, as their impact on the appearance of the bridge is significant, but on the other hand, the contractor needed to optimize reinforcement and concrete quantities since construction of the pylons was on the critical path.

The pylon shaft reinforcement layouts were developed closely with the contractor's pylon engineers and a prefabricated cage design was developed to suit the equipment and programme requirements. In the early stages of construction, a complete full-scale mock-up of typical pylon reinforcement lifts was constructed to check buildability and steel fixing times.

The cable anchorages are located within the pylon and bear against the inclined ends of the precast anchorage tie beams which resist the cable splitting forces. Because the cables are inclined both transversely and longitudinally, very accurate setting out of steel tubes in the formwork of the pylon walls was required. The Macalloy bars, which provide the necessary prestress in the anchorage tie beams, have external anchorages visible on the outside of the pylon between the exit points of the cables. They are protected by a wax blocking medium, a petrolatum compound formulated for hot injection into voids between tendons and ducts to exclude air and water and to protect the steel.

The vertical post-tensioning in the upper section of the pylon is located within the pylon as external unbonded 19T15 tendons sheathed in high-density polyethylene (HDPE) ducts, with a wax blocking medium. The use of post-tensioning reduced vertical reinforcement requirements in the highly stressed region around the cable anchor zone. The post-tensioning was required primarily to resist the bending moments introduced by the cable forces, particularly from those cables at the ends of the backspans, due to permanent loads. These forces were carefully selected to reduce moments in the backspan deck to acceptable levels.

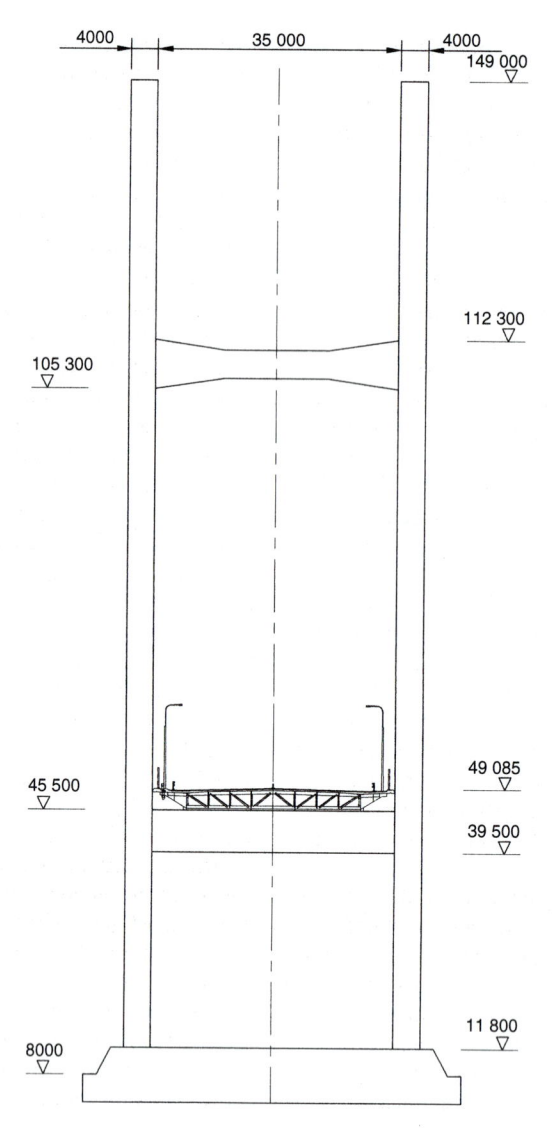

Fig. 2. General arrangement of the pylon (all dimensions and elevations in mm)

Two- and three-dimensional analyses were carried out on the pylon as part of the overall analysis of the structure. Large-displacement effects were considered using a non-linear analysis for geometric and material behaviour. Local finite element analyses were carried out in the cable anchorage regions around the precast tie beams.

Design of deck and cable stays

The layout at road level requires the deck to be slightly wider than the viaducts at 34·6 m, since the cable stays pass through the deck and are anchored to the steelwork below. A reinforced concrete slab acts compositely with two 2·15 m deep longitudinal steel plate girders and transverse members at approximately 3·6 m centres (Fig. 3). These are generally open trusses but solid plate girders were necessary at the backspan piers and at the connections with the viaducts.

The reinforced concrete slab varies in thickness between 200 mm and 350 mm over the main girders and is 470 mm thick in the anchorage regions. The 70 MPa concrete deck was cast onto the steelwork on land, leaving 2 m wide in situ sections to be cast after erection and bolting-up of each unit. Composite action is ensured by 25 mm dia. shear studs. High-strength friction grip (HSFG) bolts were used for all site connections to avoid site welding.

The cable-stay spacing along the bridge was optimized at approximately 7·3 m, which dictated the transverse girder spacing, cable anchorages being located centrally between two cross-beams within a steel fabricated assembly located on the outside of the deck between two transverse girders.

The tender design had a continuous longitudinal concrete edge beam in which the anchorages were located. It was found that an edge beam was only feasible structurally if fully continuous and reasonably deep, which conflicted with the aerodynamic requirement for narrow edge details. Design development led to the replacement of this by the intermittent steel anchorage assemblies adopted. The tender design had the main longitudinal beams quite significantly inboard and this exhibited excellent aerodynamic performance in the early wind tunnel tests but proved to be structurally uneconomical. Wind tunnel testing was programmed to run in parallel with the early conceptual structural design development and led to the adopted design with the main girders 25·2 m apart, a compromise between aerodynamic and structural requirements.

The main plate girders are generally made of steel to BS 4360 Grades 50D and DD with the heaviest bottom flanges, of size 2328 mm x 97 mm, over the backspan piers. Generally, flange sizes are smaller, varying from 750 mm x 50 mm to 1200 mm x 80 mm. The web thickness varies from 20 mm to 25 mm, increasing to 35 mm and 40 mm in the regions of the backspan piers and the pylons respectively. Rolled angle sections provide longitudinal stiffness for the webs.

The arrangement of the truss diagonals was partly dictated by the need to provide a rectangular space along the centre of the bridge for the monorail suspended access train. The truss members consist generally of fabricated T-beams, I-beams and cruciform sections made up of rolled angles and battens.

The cable anchorage assemblies transmit the longitudinal component of cable force to the deck slab by shear connections located on both sides of the two vertical web plates. Owing to their

Fig. 3. Cable-stayed bridge deck

complex geometry these units were the most difficult elements to fabricate.

In keeping with current cable-stayed bridge design philosophy, it was a Government requirement that the bridge should continue to function satisfactorily with any one cable removed for replacement. Software was developed by the designers specifically for this structure, using influence factors related to the removal of each cable.

In addition, it was required to accommodate the effects of any two cables being removed simultaneously in combination with 10% live loading. The dynamic effects of a sudden cable breakage were also considered. These cable-out load cases, and the desire to avoid large cables and to limit deck unit size led to the relatively close cable spacing of approximately 7·3 m.

The cable-out load cases, the number of cables and loading configurations, and differing combinations for ultimate load state (ULS) and service load state (SLS) led to the development of software as a post-processor to the linear elastic program LEAP 5 to provide an envelope of load effects on the principal members. For global analysis, a two-dimensional model was adequate. However, a three-dimensional model was needed to check certain locations for transverse analysis and out-of-plane behaviour, especially at the pylons.

Local effects were examined using finite element models for selected areas such as the anchorages

Fig. 4. View of cable stays and secondary cables

and to study the behaviour of the relatively thin deck slab under combined axial load, bending and shear at ULS, considering out-of-plane buckling for global and local effects simultaneously. A large-displacement analysis considering elasto-plastic material behaviour was used for this.

The whole structure is stiffened by providing two sets of backspan piers which were constructed using precast units of the same basic form as used for the viaduct piers. The deck is stressed down to the piers by four cables anchored to the deck steelwork and to a reinforced concrete slab 25 m down inside each pier. The prestress forces in these cables are sufficient to ensure that uplift cannot occur at the backspan bearings under any SLS load combinations although some separation of the bearing under extreme ULS load combinations due to extension of the tie-down cables was considered acceptable. The tie-down cables were prefabricated off site and installed as complete units. They consist of galvanized button-headed wire cables housed in HDPE ducts with a wax blocking medium for corrosion protection.

The main cable stays consist of a number of parallel seven-wire galvanized strands with tapered wedge anchorages located in wax-filled HDPE ducts, with sizes ranging from 19 to 75 strands. The number of strands was calculated on the basis of ULS loads and the characteristic breaking force of each cable. Representative cables were subjected to fatigue testing for 2 million cycles at a stress range of 162 MPa and subsequently tested to failure.

It was a recommendation of the project architect, supported strongly by the Royal Fine Arts Commission, that the cables should have a light-coloured appearance. Following investigations into the use of pigmented HDPE and the effects of ultraviolet light on the material, it was decided to provide the pale green colour chosen by the architect by applying adhesive plastic sheeting to the black HDPE sheathing prior to installation.

In the final months of construction, under certain critical conditions of wind with rain, some incidences of cable oscillations were experienced. These were generally in plane and involved the first six modes of the cables. On one occasion, some particularly violent cable motions were observed. Although mainly in plane, some out-of-plane oscillations approached ±1 m amplitude.

As a result of the movements, it was decided to fit secondary cables, or 'aiguilles', to link the cables in plane. Five such cables were fitted to each group of 30 cables (Fig. 4). With prompt action by all concerned, it was possible to design, manufacture and fit these in advance of the bridge opening.

A series of dynamic analyses was necessary to examine the effectiveness of these secondary cables, particularly because their elasticity results in non-rigid nodes where they are clamped to the cable stays and the effective increase in natural frequency of the main cables was less than that calculated on the basis of rigid nodes. These analyses modelled the main cables with the five aiguilles anchored to the deck and connected to each main cable. The aiguilles system comprised two seven-wire strands within black HDPE sheathing, passing each side of the cables and connected to the HDPE duct by purpose-made clamps. At their lower end, they are fixed positively to the deck by anchorages and tensioned. The aiguilles were installed by abseilers working from the main cables.

When the aerodynamic model test programme was devised, the feasibility of tests to ascertain cable oscillations was discussed. It was concluded that the complex conditions which combine to produce wind/rain-induced oscillations could not be modelled reliably in the wind tunnel, owing mainly to scaling problems and the variety of conditions which combine to cause this.

Simple calculations showed that cable oscillations could occur, but this would depend on the precise combination of wind speed and direction, the natural frequency of the cable, the nature and direction of rainfall, the surface characteristics of the cable sheathing, etc. In addition, there were no records of cable oscillation problems with cables encased in wax-filled ducts, although some reports had been published of oscillations with other cable forms. Although no measures for counteracting cable oscillations were included in the design, it was realized that some measures might be needed if oscillations became apparent during construction. The addition of aiguilles proved to be straightforward using abseiling techniques. In recent years an increasing number of cable-stayed bridges have had cable oscillations, often of large magnitude.

The monitoring of the dynamic behaviours of the bridge by Bristol University included some measurements of cable damping factors. These showed that the cables exhibit very low damping characteristics owing partly to the use of parallel strands, which do not lose as much energy in friction as wire ropes.

It is concluded that the damping factors used in aerodynamic model tests for cable-stayed bridges should be low enough to represent the reduced energy absorption aspects of stranded cables, if these are proposed for the completed structure.

Aerodynamic model testing

A programme of aerodynamic model tests was carried out during the design development period. Initially it involved sectional model tests, based on the central section of the main span, in both smooth and simulated turbulent winds in order to optimize the cross-section of the deck. It was decided to abandon the first sectional model based on the tender design and to replace it by one with variable transverse locations for the main girders.

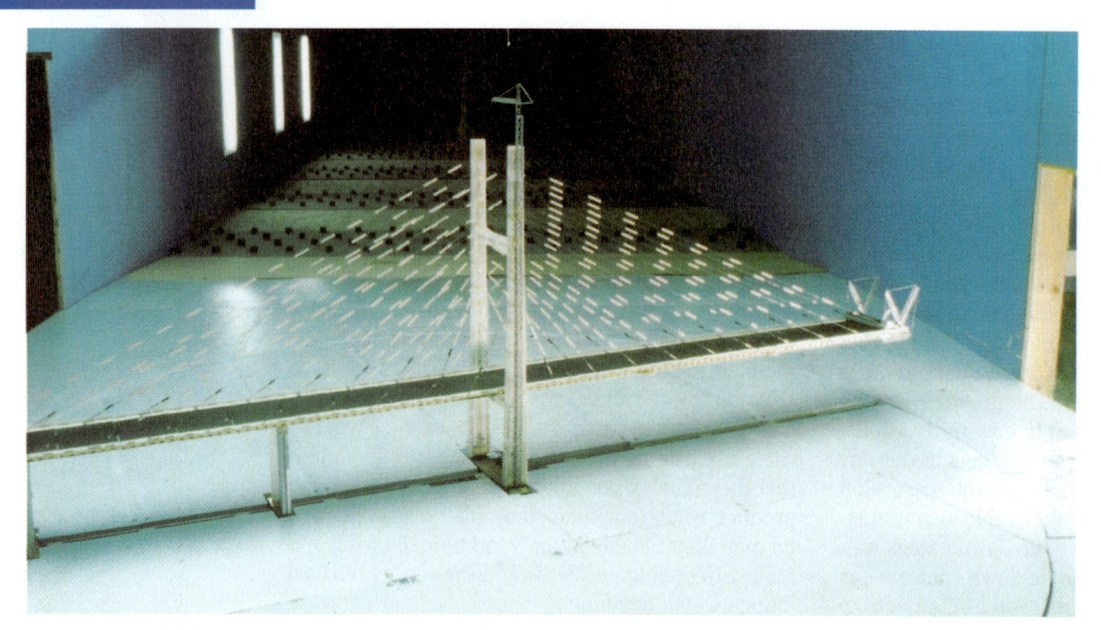

Fig. 5. Aerodynamic testing

Much experimentation was necessary before a deck configuration was found which satisfied the structural design requirements and complied with the aerodynamic criteria. As the design was developed further, a full aero-elastic model of the whole structure was tested which confirmed the earlier sectional model predictions (Fig. 5). The model was progressively dismantled and used to investigate the loading conditions and aerodynamic stability during the various construction phases. Details of the aerodynamic model tests are given by Irwin *et al.*[2]

The aerodynamic modelling was based on dynamic properties (natural frequencies and mode shapes) calculated by the designer from a three-dimensional model of the structure. The frequency ratio for the fundamental bending and torsional modes was 1·42. Measurements of natural frequency on the completed structure carried out by Bristol University showed excellent agreement with theoretical values.

Cable forces and geometry

Cable forces for the permanent load condition were selected to accommodate dead loads in addition to 'lack of fit' or 'prestress' forces which were introduced to improve the load distribution in the deck. The shop fabrication of the deck was calculated from this data to ensure that the completed deck would have the correct vertical alignment once all creep and shrinkage had taken place.

Deck erection and cable installation must be controlled on cable-stayed structures so that the correct structure geometry is achieved together with the correct distribution of permanent loads in the structure when completed. If prefabricated cables are used with their unstressed lengths known precisely and the deck units and pylons are fabricated and erected accurately this can be guaranteed. This method of cable force and geometric control is known generally as the 'cable length method'. Using this method with strand-by-strand cable installation it is necessary to measure precisely the unstressed length of at least one strand in each cable. Although this was done for a number of cables on the bridge, it proved to be impractical and was abandoned.

Instead, the designer carried out 'unbuild and build' analyses giving theoretical cable forces and structure geometry at each stage of construction and this was used on site as one means of cable force and geometry control. All conditions of the structure at each construction stage—such as deck loads and temperatures of different parts—had to be input. In parallel with this means of control, the 'reference tension' method was used and the results of both methods compared at each deck construction stage.

The reference tension method requires knowledge of the response of the deck and the pylon to an increment of cable force, measured by precise surveying. From this elastic response data for the structure, the target cable force can be calculated to ensure that the final cable force at the end of construction is correct. This method does not require knowledge of the actual loading conditions on the structure at each stage of construction. Independent teams on site calculated the target cable force and deck levels for each construction stage using the two methods and the results were compared. In general, very good agreement was reached. These means of controlling cable forces were adapted for use with the cable-stay subcontractor's method of installing the cables strand-by-strand using the Isotension method described later in this paper.

In addition to cable forces, the deck was surveyed precisely at each erection stage and adjustments were made to the target cable forces if necessary to achieve the correct deck geometry.

CABLE-STAYED BRIDGE

Construction
Pylons

Following the general philosophy of reducing work in the estuary to a minimum, pylon reinforcement cages, lower and upper cross-beams and the pylon anchorage tie beams were all prefabricated on shore.

Both the lower and upper cross-beams were constructed on raised soffits supported on concrete blocks to allow the installation of lifting beams. Once complete, the cross-beams were lifted using strand jacks (Fig. 6), placed on Lampson crawler units, as used for the caissons, and transported to the *Sar 3* barge.

The cross beams were transported to the pylons and lifted into place using the twin 1250 t capacity Lampson Transilift cranes mounted on the *Lisa A* jack-up barge (Fig. 7). The lower cross-beam weighed 1400 t and was lifted into position on the partially constructed pylons by the *Lisa A* cranes and subsequently cast into the in situ walls of the pylon. The upper cross-beam was lifted onto the top of the lower cross-beam as the *Lisa A* cranes did not have sufficient reach to install the 900 t cross-beam 110 m above the river bed (Fig. 8). The upper cross-beam was lifted into position as the pylons neared completion using two pairs of 300 t strand jacks mounted on brackets fixed to the pylons (Fig. 9).

The pylon legs are 137·2 m high above the caisson with a top level of 149 m above Ordnance Datum (OD). Each pylon consisted of 38 lifts which varied between 2·5 m and 4·48 m in height. The pylon construction started at 11·8 m above OD at the top of the caisson. The lower cross-beam was cast into the pylon legs at between 39·5 m and 45·5 m above OD. The upper cross-beam was cast in between 105·3 m and 112·3 m above OD (Fig. 2).

The pylons were constructed by traditional lift over lift methods using a Peri ACS self-climbing formwork system, which was adopted because of the expected tower crane down time caused by the high winds in the estuary. The six working platform levels on the Peri system provided access to all reinforcing, formwork fixing, concreting and formwork climbing operations plus a landing platform for the temporary Alimak hoists

Fig. 6 (top). Upper cross-beam being lifted by strand jacks

Fig. 7. Lifting of lower cross-beam

55

which operated on the outside face of each pylon to provide access for men and materials (Fig. 9).

The concrete for the pylon in situ construction was batched by a 40 m^3/h capacity Steelfields batching plant mounted on top of the caisson foundation and placed by skip using the Liebherr 290 HC tower cranes which were anchored to the pylon and climbed as the pylon was constructed. The area on top of the caisson foundation was also used for cement, storage of ground granulated blast furnace slag (GGBFS) and aggregate, messing facilities for the workforce and office accommodation for the engineers and supervisory staff.

The construction of the pylons became particularly difficult at the lower cross-beam level where the structural connection into the pylons required a substantial increase in the density of reinforcement. The complexity of pylon construction was greatest in the region where the cable stays were anchored into each of the pylon faces perpendicular to the line of the deck, owing to the presence of the steel formwork tubes for the cable stays, the tie beams and the high density of reinforcement.

Because of the geometry of the deck and the pylon, the cable stays are not in a vertical plane and each anchorage has a unique orientation. The difficulties of casting the anchorages in situ to the close tolerances required led to the decision to precast the anchorage tie beams on shore. The beams incorporated a pair of cable-stay formwork tubes and anchor plates plus polyvinyl chloride (PVC) sleeves for the horizontal Macalloy bars. The close tolerances for position and vertical and horizontal alignment of the anchorages were more easily controlled in purpose-made formwork jigs on shore (Fig. 10). The completed beams were lifted into position by the tower crane and set on jacks placed on the tie beam below. The reinforcement and formwork for the four walls of the pylon shaft were fixed around the anchorage beam prior to concreting.

When pylon construction had progressed above the level of the upper cross-beam a pair of temporary steel lifting brackets was attached to the pylon walls and the strand jacks installed on the brackets. Prior to lifting the upper cross-beam into position the brackets and strand jacks were subjected to a 125% load test, by using the upper cross-beam anchored to the top of the lower cross-beam as a test weight. The strand jacks hoisted the upper cross-beam 60 m to its permanent position at an approximate rate of 30 m/h (Fig. 9). The deflection of the pylon legs was measured during the lifting operation, and prior to the final positioning of the beam the legs were jacked back to their undeflected alignment.

Once in position, an in situ stitch between the pylon and the upper cross-beam was made before the upper cross-beam was post-tensioned onto the pylons using seven pairs of 37T15 stressing cables which were grouted into the structure after stressing.

Following the completion of both pylon legs, 16 vertical tendons consisting of 19T15 strands were installed in HDPE ducts in each pylon leg and tensioned. The cables were anchored below the level of the first cable stay and in a slab at the top of the pylon.

Survey control of the pylons during construction was important as the initial alignment was designed to accommodate the deflections expected during the deck erection phase.

The programme dictated that both pylons were under construction at the same time, necessitating two complete sets of pylon construction equipment.

As the centre of the bridge was in excess of 2·5 km from the Avon shore, the access to both pylons required careful planning. The Avon pylon was connected to shore at low tide by the causeway and the Gwent pylon could be accessed by boat across The Shoots channel. Both the Avon and the Gwent pylons were accessed at high tide by boat. All of the materials for the construction of the pylons including cement, GGBFS, sand, aggregate and reinforcement were transported either by barge at high tide or by way of the causeway at low tide, the latter for the Avon pylon only. To aid the transport of personnel to the

Fig. 8. Lifting of upper cross-beam

CABLE-STAYED BRIDGE

Gwent pylon by boat a small jetty was constructed on the Gwent shore.

Deck and cable stays

The ethos of constructing as much of the bridge on shore as possible was also adopted for the composite cable-stayed bridge deck and a purpose-built production yard was established for the assembly and casting of the deck units. The deck units were then transported at low tide on a multi-wheeled trailer by way of the roll on–roll off jetty to the *Sar 3* barge. At high tide the *Sar 3* transported the deck unit into the estuary where the double-shearleg (DSL) cranes lifted it into position.

The cable-stayed bridge deck was constructed by the cantilever method from each pylon, the Avon side being constructed first, followed by the Gwent side. The joints between the structural steelwork longitudinal beams were HSFG bolted and the stitch between the precast deck units was cast in situ. Following placing of the concrete stitch, the cable stays supporting the deck were installed and stressed.

Exceptions to the standard deck erection cycle occurred at the units on the pylons, the units on the four backspan piers and the units connecting the cable-stayed bridge to the viaducts. The erection cycle around the backspan piers was complicated by the need to provide a tie-down force during deck erection. The stressing sequence of the cable stays was also more complicated around the backspan piers where some retensioning/detensioning was required.

The structural steelwork was fabricated at the Cimolai factory in Italy and trial-erected to ensure that the correct geometry had been achieved.

The units were then shipped to Avonmouth Docks and brought to site by road in six pieces

- two longitudinal plate girders complete with transverse cantilevers and the cable-stay anchorage beam
- four transverse truss sections which bolted together in pairs to connect the two longitudinal plate girders.

A production yard was constructed on shore which consisted of two pairs of parallel rails on ground bearing reinforced concrete beams, capable of supporting 650 t and 250 t units respectively. The units on the pylons, backspan piers and backspan units were generally constructed on the 650 t line and the lighter, centre-span units were produced on the 250 t line.

The match-casting philosophy adopted for the project as a whole was also applied to the structural steelwork for the cable-stayed bridge deck. To achieve this, each deck unit being concreted had the previous and subsequent deck units attached to it to maintain the set alignment. To control the vertical alignment to the predetermined profile, the deck units were supported on hydraulic jacks. Once the alignment had been achieved, it was maintained by reaming out the corner holes of the vertical web splice plates of the longitudinal girders of two adjacent units and inserting close-tolerance pins into these holes. These maintained the alignment while the units were being winched along the production line and during all subsequent construction operations.

The original plan had been to drill the splice plates for the web and the flange and one end of each girder in the factory. Following the align-

Fig. 9 (left). Upper cross-beam lifting brackets and pylon-climbing formwork

Fig. 10. Precast cable anchorage tie beams

ment of the girders on the assembly line, the remaining holes would have been drilled using the splice plates as a template. Cimolai were so confident they could achieve the required tolerances that they drilled all the holes in the factory. Every splice was made within tolerance, with significant savings to the programme of work on site.

The support rails in the fabrication yard were set at the same centres as the longitudinal girders of the bridge deck. The steelwork was on machinery skates running on the support rails and was controlled by regular survey to ensure that the close tolerances specified were achieved. Once the steelwork had been bolted together it was winched along the rails to allow a further unit to be assembled behind it. The assembled steelwork was positioned over a specially designed formwork and falsework system which provided the soffit for the deck slab. Once the formwork had been raised, the deck slab reinforcement was fixed and the concrete placed. The concrete strength gain was carefully monitored by maturity measurements and the deck soffit formwork and falsework was struck as soon as the concrete was capable of supporting the loads applied.

Following the striking of the soffit formwork the concreted unit and the assembled steelwork attached to it were winched forward to allow the assembly of the next deck unit at the start of the production line. Each deck unit slab was cast while attached to a completed unit in front and an assembled steelwork unit behind. Following the concreting of the second deck unit in the line and the subsequent winching forward of the three deck units, the lead unit was disconnected and winched forward to the end of the production line, where the unit was weighed before being transported by an Econofreight multi-wheeled trailer to a purpose-built enclosed painting shed. The deck weights were an essential requirement of the deck alignment control system and accurate weighing of the deck units at the end of the production line was achieved using load cells calibrated to within 2%.

The steel was delivered to site with the first two coats of epoxy blast primer and micaceous iron oxide having been applied in the factory. The final coat of polyurethane paint was applied on site after the unit had been assembled and the in situ concrete placed. The painting area was enclosed and heated to maintain the environment required for successful paint application. The Control of Substances Hazardous to Health (COSHH) requirements of the paint were managed by the installation of a forced ventilation system to allow safe application.

An Econofreight multi-wheeled trailer was used to transport the completed deck unit to a storage area where final fitting of ancillary equipment was carried out. The deck units were kept in storage until the erection programme required their transportation to the cantilever under construction (Fig. 11).

Owing to programme constraints a number of the heaviest deck units, for example the unit on the Gwent pylon and the four units for the backspan piers, were cast out of sequence in a separate area of the construction yard (Fig. 12). The same principle of casting the units attached to the previous and subsequent units in the deck sequence was followed, to maintain the vertical alignment but without the benefit of the support rails. The leading and trailing deck units to the special deck units were reintroduced into the construction sequence on the production lines to maintain the vertical alignment of the whole deck.

The completed deck units were transported individually to the pylon by the *Sar 3* motorized barge. The initial unit on the pylon was lifted into position on the lower cross-beam by the two 1250 t capacity Lampson Transilift cranes mounted on the deck of the *Lisa A* jack-up barge. The *Sar 3* was controlled by laser and computer technology which enabled the barge to maintain its position within 500 mm of

Fig. 11 (top). Standard deck unit on the Econofreight transporter

Fig. 12 (above). Pylon deck segment being lifted in the yard

CABLE-STAYED BRIDGE

the intended position while the tide was running at approximately 2–3 knots. In the Severn estuary the tide could run at up to 10 knots; therefore, the transportation and lifting operations were limited to 1 h each side of high tide.

The unit on the pylon was placed on temporary bearings on the lower cross-beam and temporarily stressed down to the beam using Macalloy bars (Figs 13 and 14). One pair of standard units CO1 and BO1 was erected on each side of the central unit on the pylon using the *Lisa A*, to create enough room on the deck to allow the installation of DSL cranes by the *Lisa A*.

The first cable stays were attached to the CO1 and BO1 deck units prior to the erection of the DSL cranes, which were used to erect the standard deck units in the cable-stayed bridge centre span and backspan. They were pairs of luffing cranes mounted on a common body which was anchored through the deck to the cantilevering structural steelwork and had the capacity to lift 289 t at 9·6 m radius and 236 t at 16·4 m radius (Fig. 15). The cranes followed the leading edge of the construction by skidding along channels placed on the deck using hydraulic rams fitted to the crane frame. The DSL cranes had no capability to slew and hence were positioned on the deck accurately, but more importantly the delivery of the deck unit had to be controlled to a highly accurate position on the river 50 m below deck level. Both of the cranes were controlled by a single operator who supervised the lifting operation from deck level (Fig. 16).

The lifting frame was permanently attached to the DSL cranes. The deck unit was most at risk while the lifting frame was being connected to the

Fig. 13. Transportation of the segment to the pylon

Fig. 14 (above). Lifting the segment onto the pylon

Fig. 15 (top right). DSL cranes in the yard awaiting transportation to the bridge

Fig. 16 (right). DSL cranes lifting a backspan unit on the Avon deck

deck unit and the *Sar 3* barge was holding position under computer control. A simple pinned connection for the lift was achieved by bolting brackets to the cantilevering structural steelwork while the unit was on shore. A three-point lift of the deck unit was ensured by incorporating hydraulic rams into the system to transfer the deck weight to the lifting frame.

The standard erection cycle commenced with the CO2 unit which was transported to the pylon by the *Sar 3* barge. The centre-span DSL crane was used to pick up the unit from the barge deck and hoist it to deck level. The unit was lined up and the four close-tolerance pins used in the deck construction yard to control the vertical alignment were reinstalled through the splice plates making up the longitudinal girder web joints. The deck unit and the unit previously erected were then surveyed and a predicted angle between the units was set.

With the angle between the units set, the HSFG bolts for the longitudinal girder joints were all inserted through the splice plates and checked to ensure that they were all free before being tensioned. Load-indicating washers were used to control the tension. The alignment of the deck unit was checked after bolting and preparations were made to concrete the in situ 2 m wide stitch between deck units. Access for bolting the joints

CABLE-STAYED BRIDGE

between the longitudinal girders was provided by a platform suspended from the permanent gantry rails attached to the transverse trusses. The access gantry also transported the formwork for the in situ stitch which was raised to the underside of the two adjacent deck units and hung from ties passing through soldiers spanning across the gap.

While the deck slab reinforcement was being fixed, an epoxy grout was poured into the 50 mm gap between a series of nibs precast on the edge of each deck unit. Once the epoxy grout had reached the specified strength, the horizontal loads generated by the cable stays could be transferred across the joint by way of the nibs, therefore removing the slower strength gain of the stitch concrete from the critical path of the erection cycle. The weight of the concrete was still required to balance the deck but as soon as the in situ stitch had been poured the cable-stay installation could begin.

PSC Freyssinet used a strand-by-strand installation system to construct each cable. The individual strands were fixed into the anchorage by a system of wedges which gripped the tensioned strand and locked against tapered holes machined in the anchorage plate.

Initially, the cable-stay sheath and the first or reference strand were laid out on the deck. The end of the cable nearest the pylon was picked up by one of the tower cranes attached to the pylon (Fig. 17). The opposite end of the cable strand was threaded through the lower anchorage, a load cell and the monostrand stressing jack. The upper end of the strand was threaded through the anchorage inside the pylon and wedged into place. The monostrand jack applied a tension to the reference strand equal to approximately 60% of the final tension.

As the jack applied tension to the cable it also forced the wedges into the anchorage. A second strand was then threaded through the lower anchorage up the cable-stay sheath and through the upper anchorage, where it was wedged into place. The second strand was tensioned using the monostrand jack until the load measured by the load cell on the reference strand and the load in the second strand were equal, at which stage the computer controlling the monostrand jack stopped the stressing operation (Fig. 18). The same process, known as the Isotension system, continued in the pair of cables supporting the deck unit until all the strands in the cable had been installed and stressed to 60% of the final cable load. At this stage the deck and pylon were surveyed to assess how the structure had reacted to the applied load.

The actual alignment of the deck was compared to the theoretical alignment and an adjustment was made to the force to be applied to the cable to achieve the 100% cable load. The cable force at each construction stage was calculated using the reference tension method and the force method

Fig. 17. Cable-stay installation

Fig. 18. Isotension stressing of the cable stays

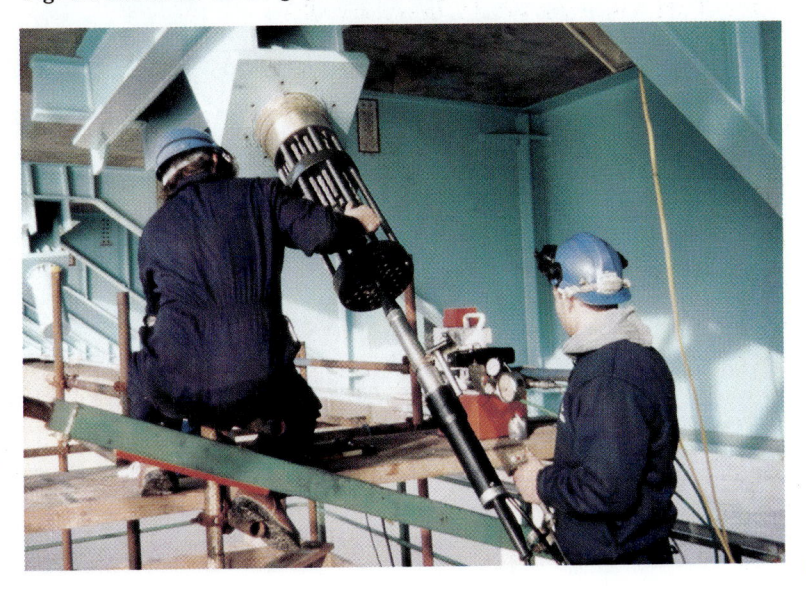

described above. As the load applied to the strands could be controlled more accurately by extension than by load measurement, the final load increment was applied by extending each strand by the same amount. After final tensioning, the centre-span and backspan decks were surveyed again so that the angles for the next units could be assessed to meet the vertical alignment of the whole deck. The same cycle was followed for the backspan unit so that the deck cantilever remained in balance.

The deck erection cycle continued until the first backspan pier was reached. The weight of these units was beyond the capacity of the DSL cranes and, in addition, the DSL was incapable of reaching beyond the caisson to pick up the next unit from the *Sar 3* deck. Therefore, it was decided to cast both units together and lift the pair using the Transilift cranes on the *Lisa A*. Once the backspan deck units had been installed in the deck erection sequence the eight vertical tie-down cable stays were installed.

After the backspan pier unit had been erected the standard erection cycle was continued up to the next backspan pier, when the above sequence was followed (Fig. 19).

Following the erection of all the units from the Avon pylon, the centre-span DSL was skidded back along the deck to the end of the backspan. Both DSLs were lifted from the deck and relocated on the Gwent pylon units to begin

Fig. 19. General view of balanced cantilever construction

Fig. 20. Last few segments to be placed on the Gwent span

the erection cycle (Fig. 20).

The final unit to be placed in the centre span was a standard unit. After placing the penultimate unit on the Gwent side, the two ends were accurately surveyed and after making temperature corrections, the gap was found to be correct within 20 mm but with different gaps between each girder. The splice plates were then drilled to suit to correct for the differences.

During the erection of the balanced cantilevers, the decks were effectively fixed longitudinally to the pylons by modifying the hydraulic circuit on the shock transmission units to prevent any movement. By pumping hydraulic fluid into one side of each unit, it was possible to move the decks apart to allow sufficient clearance to erect the last unit, which was lifted and connected to the Gwent cantilever in the normal manner.

Following the stressing of the last cable stays, there was still a significant level difference between the two cantilevers because of the weight of the DSL. The DSL was skidded back towards the pylon and kentledge was added to the Avon cantilever to equalize the levels. The shock transmission units were used to move the decks longitudinally to enable the girders to be spliced together. The hydraulics were then modified to allow the units to perform as intended for the permanent condition, as combined springs and shock absorbers. The in situ stitch was concreted to complete the main span.

The closure of the backspan deck (Fig. 21) onto the viaducts is described by Mizon and Kitchener.[3] Since there was a movement joint and a physical gap between the ends of the two structures, there was no need to use the shock transmission units to allow the unit to be erected.

On completion of the deck erection, the DSLs were stripped down on the deck and returned to shore to allow the deck furniture and surfacing to be completed.

Acknowledgements

The authors acknowledge with thanks the assistance received from many of their colleagues in the preparation of this paper and in particular C. M. Tong of Halcrow-SEEE and J. N. Kitchener of Laing-GTM. Photographs are reproduced with kind permission of Photographic Engineering Services.

References

1. CONCRETE SOCIETY. *Recommendations.* July/August 1993.
2. IRWIN P., MIZON D., MAURY Y. and SCHMITT J. History of the aerodynamic investigations for second Severn crossing. *Int. Conf. AIPC-FIP, Deauville, France,* 1994.
3. MIZON D. H. and KITCHENER J. N. Second Severn crossing—viaduct superstructure and piers. *Proc. Instn Civ. Engrs, Civ. Engng, Second Severn Crossing,* 1997, 35–48.

Fig. 21. Connecting the cable-stayed bridge and the viaduct

KITCHENER
AND MIZON

Second Severn crossing—management of construction

J. N. Kitchener, BSc, CEng, FICE, *and D. H. Mizon*, CEng, MICE, MIStructE, MIHT, MIL

Construction of the second motorway bridge over the Severn estuary between England and Wales was another triumph of Anglo-French collaboration. Management of the £330 million project was split equally between the British and French construction partners and resulted in completion on time, within budget and without loss of life or limb. This paper describes the management strategy adopted for the project, explains the role of the designer and Government's agent during construction and reports on concrete production, quality assurance, safety, environmental, human resources and community relations issues.

Management strategy

The two companies Laing and GTM, which led the winning bid for the design, build, finance and operate concession of the second Severn crossing between England and Wales, formed a fully integrated joint venture for the construction work. A board of five directors from each company was set up to oversee the contract and regular board meetings were held on site to review the status of the project and to make overall policy decisions. The day-to-day technical, commercial and construction matters were left to the British project director and the French project manager, who were resident on site throughout the contract period.

A headquarters team comprising the project director, project manager, commercial manager, chief engineer, construction manager and the heads of the various service departments was established on site.

The upper levels of the construction staff hierarchy were evenly split between British and French staff but the lower levels were predominantly British. At peak there were over 200 staff seconded to the joint venture, of whom 40 were French.

In order to control the construction of the works effectively, the contract was split initially into four sections.

- *Avon*—responsible for all the permanent works on the English shore, the construction of the precast units for the caissons and piers and the manufacture and erection of the Avon viaduct deck.
- *Gwent*—responsible for all the permanent works on the Welsh shore, including the toll plaza and the manufacture and erection of the Gwent viaduct deck.
- *Marine*—responsible for all marine construction works including the placing and concreting of the caissons, the marine piled foundations, the erection of the precast piers, all heavy lifting operations, and the operation of the extensive marine fleet and the causeway to provide access around the river for men and materials.
- *Main bridge*—responsible for the construction of the cable-stayed bridge, including the prefabrication of the deck segments on shore.

A fifth section, concerned with all the finishing operations, including surfacing, carriageway furniture, services and all mechanical and electrical works, was established at a later date. Each section was managed by a section manager as an autonomous unit with its own dedicated team of supervisors, engineers, planners, quantity surveyors and workforce, although in technical matters, each service department also reported to its head of department. Other service departments such as safety, quality assurance (QA), plant, industrial relations, purchasing, accounts, cashiering and public relations were organized centrally by the headquarters team.

The project team, comprising the five section managers and the heads of the various service departments, reported directly to the project manager or director. They were responsible for overall coordination and ensuring the technical and financial competence of the project as a whole.

Role of the designer on site

The contract required that the designer was represented on site during construction by a team led by a designer's representative (DR) who had

Neil Kitchener, chief engineer at Laing-GTM

David Mizon, a design manager at Halcrow-SEEE

MANAGEMENT OF CONSTRUCTION

been closely involved in the design of the crossing. The principal function of the DR's team was to ensure that the crossing was built as the designer intended, in accordance with the relevant contract documents, and to certify this for each element of the construction.

The contractor's quality system was devised with this in mind and was written to ensure that the DR played an active role in all construction and certification processes. Each construction activity was documented by a series of quality documents such as technical queries, compliance appraisal forms, method statements, inspection reports, non-conformance reports, etc. These were originated by the contractor and submitted to the DR throughout each construction stage, further construction being prohibited unless the requisite forms were signed. A fully operational system involving the DR at all essential stages, based on the contractor's previous recent experience on other major projects, was implemented from the outset of the project. This proved to be a distinct advantage to all parties involved. To operate effectively within the QA framework and to discharge his contractual obligations, the DR had an Anglo-French team of 20 engineers, technicians and support staff. This provided excellent opportunities for French engineers to gain valuable experience in Britain on a large project. Also, the cross-fertilization of ideas, cultures and concepts broadened the knowledge and vision of all members of the team.

In addition to the site surveillance and certification role, the DR's team was readily available to assist and advise the contractor on construction methods and sequences, temporary works, loading criteria and so on using its knowledge of the bridge design and loading criteria. On such a technically demanding project, a large proportion of the DR's input was spent on these technical matters, in particular examining the effects of construction methods and equipment on the permanent works. Examples of this were caisson lifting, weighing and placing, cable-stay force calculations for each erection stage, viaduct deck jacking calculations and concrete mix design. In addition, the DR's geotechnical engineers and geologist inspected each foundation site before any construction started and each caisson founding level was reviewed together with the contractor's geotechnical engineer. Also, pile lengths were recalculated using data from trial piles and, in some cases, data from additional boreholes.

The DR's team was also jointly responsible with the contractor for preparing the project maintenance manual, which was handed over to the concessionaire at the end of construction. This included the preparation by the DR's team on site of 900 as-built drawings.

Each item of a division of the works given in the construction contract had to be certified for compliance when complete. This process required auditing all relevant quality documents and a review of all surveillance reports and any correspondence on the items in question, in addition to detailed site inspections. Each construction certificate was signed by the contractor, designer, Government's agent and concessionaire.

To assist in the DR's routine surveillance and to provide formal records, random audits or construction controls were carried out on a regular basis. Certain areas of the works were targeted for these random technical audits, where the sources of information, lines of responsibility, self-checking techniques and so on used by the contractor were scrutinized and reported on. These construction controls were outside the contractor's quality procedures and proved to be a useful independent investigation into the contractor's methodology and technical approach to each aspect of the project, especially during the earlier part of the construction programme.

The extensive inspection and maintenance facilities involved more intensified activity on mechanical and electrical (M&E) works in the latter part of construction. Specialists in these fields provided additional assistance from the designer's permanent offices when required. Each item of access equipment—such as gantries, access train and support system, pylon lifts and walkways—was tested under a variety of loading and operational conditions.

Role of the Government's agent

The Government agent, Maunsell, maintained a staff of eight on site throughout the construction contract period in order to

- approve and monitor the contractor's construction and quality assurance system
- review the contractor's design, drawings and specification for compliance
- implement the Government's changes
- assess concessionaire's changes
- evaluate contractor's and concessionaire's claims
- approve construction certificates
- review and monitor environmental protection measures
- instruct and supervise entrusted works
- inspect and accept the completed works
- review and approve maintenance manuals
- monitor reinstatement of construction yards and site
- ensure the parliamentary undertakings were fulfilled
- provide a focal point for correspondence between all parties to the agreement
- provide liaison to all other parties including approach roads contracts
- review and accept unresolved non-conformance reports
- issue the permit to use on completion of the contract.

The wording of construction certificates in the contract assumed that each item of work, often valued at several million pounds, was 100% complete and within specification when the certificate was signed by the parties, for progress and payment purposes. This was revised through the QA procedures to make each item in effect a 'substantial completion' certificate with snagging lists attached. Neither contractor, designer nor concessionaire could accept anything outside the specification, all such matters being referred to the Government's agent for approval.

The Government's changes, requested by the Government's agent, had, in theory, to be priced by the contractor before being officially processed. The contract contained no detailed schedules of rates and the only prices were often in terms of millions of pounds for 'milestone' payments. Also, since the Government's agent normally instigated such change requests as early as possible to avoid disrupting the works, the contractor invariably had no detailed prices. In practice, all changes proceeded before prices were agreed.

The Government's changes were generally issued as a result of keeping up to date with revised national specifications and technical advances during the period of up to eight years from preparation of the specification to execution.

The concessionaire's changes had to be examined to ensure that the revised proposal resulted in no less a standard of the finished product. The concession did not permit any financial benefit to the Government resulting from the concessionaire's changes. The Government's changes invariably cost the Government more.

Although some risks, including unforeseen ground conditions, were accepted by the concessionaire, there were still claims where the Government's requirements and concessionaire's proposals were not fully 'back to back' or were in other grey areas of interpretation. There was no clause which gave precedence to the Government's requirements.

The concessionaire invoked a clause delegating full authority to the contractor to negotiate construction claims directly with the Government agent. There were of course occasions where the contractor had a valid claim against the concessionaire through the construction contract but the concessionaire had no similar claim against the Government through the concession agreement, although the contractor's entitlement to payment was restricted to the amount which the concessionaire received from the Government. Nevertheless, the total cost of additional works and settled claims amounted to only approximately 1% of the value of the construction contract.

The designer's responsibility for overseeing construction did not extend to any environmental effects (noise, dust, traffic, etc.), which fell to the Government agent to monitor and control with the contractor.

Over the latter months before opening, the Government's agent's time was increasingly spent on examining and reviewing the final QA documentation packages and resolving non-conformances satisfactorily. Following issue of the overall substantial completion certificate, the Government's agent had a maximum of 45 days to carry out a final physical examination of the works before issuing the 'permit to use' which allowed the crossing to be brought into use. This final inspection, together with a summary of all changes and significant non-conformances, was produced for the Government in a pre-opening report.

The one-year maintenance period was spent in finalizing the cost of claims, Government changes, entrusted works and further inspections, as outstanding and remedial works were completed.

Quality management

The construction works were carried out under a quality assurance regime in accordance with BS EN ISO 9001: 1994 (formerly BS 5750 Part 1). The QA manager, who reported directly to the project director, ran a small team of engineers who ensured that all the necessary procedures were produced and audited all aspects of the site, including the quality systems of the designers and subcontractors. They carried out vendor assessment of potential subcontractors and suppliers as well as undertaking surveillance, inspection and final acceptance of off-site fabrications and manufactured items. The close involvement with this work ensured that delivery dates were achieved and that only materials complying with the specification were delivered to site, thus avoiding expensive and time-consuming delays to work on site which could arise from late delivery or defective materials.

In addition to operating the site documentation centre, they were responsible for the collection and collation of all documentation associated with the lifetime records packages and the maintenance manuals. They also played a key role in the progressive certification of the project by the designer, the Government's agent and the concessionaire.

The quality system which ensured that work on site complied with the contractual requirements was based on the production and approval of quality plans and method statements for each operation on site.

Each section was directly responsible for the quality of its work and most of the inspections were carried out by the site engineers. The quality control manager, with his team of up to eight inspectors, played a pivotal role in maintaining standards by undertaking random audits of the work.

Construction certification to signify satisfactory completion of construction activities was required at predetermined intervals throughout the contract period. This involved the Government's agent's approving milestone certificates endorsed

MANAGEMENT OF CONSTRUCTION

by the designer, contractor and concessionaire. On completion of each milestone activity, the certification procedure was instigated, whereby full QA documentation was made available to the designer's representative for review. In conjunction with his ongoing site surveillance inspections and construction control audits, he was able to satisfy himself that all aspects of construction were complete. The approval of a construction certificate allowed the concessionaire to release the milestone payment for that particular construction activity.

Safety

In addition to the normal risks involved on a construction site, the complex nature of the work, the large number of heavy lifting operations and the difficulties associated with marine operations combined to make safety a major issue on the project. A health and safety plan for the project was drawn up by senior management and the resident safety manager, which identified the general nature of the risks and laid down a procedure for establishing safe systems of work.

No operation could be carried out on site without an approved method statement. The risks and technical difficulties involved were evaluated and used to classify each operation as minor, normal or major. All method statements, risk assessments and procedures were subject to review by suitably qualified and experienced staff, sometimes involving off-site expertise. Although similar principles were used for preparing any method statement, the level of authority required for approval varied according to the classification and any which were classified as major could only be approved by the chief engineer. In accordance with company practice, all temporary work schemes and heavy lifting operations were also independently checked and certified. Any scheme which involved loading of the permanent works was also submitted to the designer to ensure that the structure was not compromised.

The major operations such as caisson handling and viaduct deck erection evolved over a lengthy period of time and risk management techniques were used to ensure that the concepts were correct. The basic method was decided before the site teams were fully involved and the permanent structure and the handling plant were designed accordingly. However, the site management teams were ultimately responsible for ensuring that all operations were carried out safely and they had to carry out the risk assessments and prepare the detailed method statements for each operation, using the skills and resources of each department as appropriate to the operation. As part of the approval processes, the construction teams frequently had to make detailed presentations of their plans to senior management and on occasions to the factory inspector.

During construction, all operations were constantly reviewed and changes to improve both productivity and safety were made as necessary. Although site management was responsible for safety, the safety department carried out regular inspections of work in progress and was responsible for the significant amount of site safety inductions, for safety training and for the operation of the site medical and health screening centre. Audits of the safety management system were carried out by both the site safety department and an independent safety auditor.

The early introduction of safety representatives and monthly safety committee meetings was beneficial to safety on the contract by raising the involvement of the workforce. Regular toolbox talks were used as the vehicle for maintaining the awareness of all operatives for routine matters, and briefing sessions were organized for those involved in specific operations.

Close and regular liaison between the site team and the Health & Safety Executive was encouraged and helped to foster a significant improvement in the understanding of each other's position and the problems of the construction.

A heavy lifting team, which included experienced crane supervisors, operators, riggers and engineers was set up under the control of the marine section manager. They carried out every major operation involving lifting loads from the barges and assisted in many of the complicated lifts on shore using cranes and jacking equipment.

The work in the river was perceived as one of the areas of greatest risk and a marine control room was established in an elevated position on the Avon shore with a good view over the estuary. It was manned 24 h a day by experienced radio operators and maintained radio contact with all the boats and every gang working between the flood banks on either shore. The radio operators were in regular contact with the local weather bureau and provided a weather and tide forecasting service for the site. The marine control room also housed a large computer network and the site radio system.

In view of the vast tidal range and the nature of the river bed, it was vital to know the level of the tide at all times and an environmental monitoring station was established on one of the existing navigation beacons in the middle of the river to monitor tide level, wind speed and direction and wave height. The data was transmitted by telemetry link to the computer network in marine control where the information was permanently displayed. A back-up tide gauge was also provided because of the importance of the information.

The tide rises and falls at up to 3 m/h at spring tides and safe navigation across the estuary, full of rocky outcrops, was vital. A sophisticated computer system was set up which incorporated a digital ground model of the estuary in the area of the crossing, a tidal model based on the Admiralty tidal forecasts, and data from the hydraulic model studies concerning stream velocities and directions. Using this data, the marine supervisor was able to pre-plan any route across the estuary to ensure that the boats could complete their operation during the limited tide window and return to the jetty before the river dried out.

The major boats were also provided with on-board computer systems to aid their navigation. This relied on an on-board position fixing system which was provided either by radio location beacons accurate to 0·5 m or a real-time differential global positioning system (DGPS) which was only accurate to 10 m. The position of the boat and the actual level of the tide were fed into the computer, which calculated the depth of the water below the keel and displayed a chart of the estuary showing the areas where sufficient water was available for navigation, along with the actual position of the vessel itself in real time. The chart of the estuary was regularly updated as construction progressed and all obstructions on the bed were displayed. This simple display enabled the captain to navigate around the river, confident that there was sufficient water below the keel, and was infinitely more accurate than the radar sets which were also provided as a means of back-up and locating the other craft.

In order to safeguard personnel working on the river, a further control room was established at the entrance to the river to enforce the safety procedures. As each person crossed the sea wall, he was issued with a numbered life jacket which was used as a tally, like the miner's lamp, to provide a continuous check of the number of personnel on the river and their work location. The control room staff also controlled all vehicles which used the causeway and ensured that it was safe for use when the tide went out and that all personnel working on the river bed had returned to the shore before the tide came in.

A fleet of small safety boats was also provided, which were manned whenever tide conditions permitted and personnel were working on or above water (Fig. 1). During the construction of the main bridge, a safety boat was kept on station in the navigation channel under the main span 24 h a day. Extensive planning and numerous practice exercises were held with the local rescue and emergency authorities to ensure that rescues and evacuation could be made from anywhere on the crossing.

In order to check the effectiveness of safety systems, various measures were adopted. These included

- simple measurements of safety achievement, for example the number of people trained
- inspections of the physical conditions by supervisors, managers and the safety team
- analysis of accident data
- a numerical audit system with targeted improvements
- a quarterly independent safety audit.

Fig. 1. Marine access to the pylons

Subcontracted activity	Subcontractor
Prestressing	GTM Btp
Structural steelwork	Cimolai (UK) Ltd
Cable stays	PSC Freyssinet
Surfacing	ARC
Access train and inspection gantries	Laing-Lentjes JV
Environmental monitoring	SGS Environment

Table 1. Principal subcontracts

Despite all the efforts made to provide safe and healthy working conditions, it would be wrong to suggest total success. There were accidents and dangerous occurrences, including some serious ones. However, through the considerable efforts of everybody involved and a little bit of luck, the prime objective of completing the project without loss of life or limb was achieved.

Resources

In view of the physically and technically challenging nature of the project and the necessary degree of cooperation and coordination, it was decided that the majority of the work would be carried out by a directly employed workforce and only specialist trades were subcontracted out. At peak, over 900 operatives worked on site, with significant numbers involved in the off-site fabrication and manufacture of specialist plant and equipment, many based across Europe.

The majority of the operatives came from the large local catchment area but a significant number of the tradesmen and marine crew were trav-

MANAGEMENT OF CONSTRUCTION

Fig. 2. Marine plant near the Avon yard

elling men. No French operatives were directly employed by the joint venture but some worked for the prestressing subcontractor GTM Btp. A large amount of training was carried out on site to teach the specialist and operational skills required on the project. The major specialist works which were subcontracted are shown in Table 1. Numerous smaller subcontracts, particularly for the finishing works and specialist mechanical handling equipment, were also placed.

Concrete production

Apart from a small quantity at the beginning and end of the contract, all the concrete for the project was produced on site in one of six batching plants. This provided the necessary flexibility and coordination of supply and ensured that the quality was achieved.

All mixes, ranging from 10 MPa blinding to 70 MPa structural grades, contained ground granulated blast furnace slag to provide enhanced durability, reduce thermal problems during curing and for economy. Local coarse and fine limestone aggregates were blended with marine dredged sand from the Bristol Channel to produce an acceptable grading. Chemical admixtures were also used to improve the properties of the concrete in both the wet and the hardened state. On shore, the concrete on the Avon construction yard was batched in an Elba EMC105 plant while in Gwent, an Elba EMC60 was used. The concrete for the precast viaduct units was predominately pumped directly into the moulds using placing booms, while the remaining concrete was transported by truck mixer and subsequently placed by pump or conventional crane and skip.

The marine concrete was produced on the specially adapted jack-up barge *Karlissa B* by two Elba EMC105 plants and the concrete was pumped directly into the pours using placing booms (Fig. 2). Small quantities of concrete were transported from the *Karlissa B* in an agitator drum mounted on one of the work boats and placed by crane and skip.

The concrete for each of the pylons and the in situ deck stitches on the main bridge was produced by a Steelfields Major 60 plant sited on the top of the caisson and placed by crane and skip.

Site laboratories were set up in the Avon and Gwent construction yards with satellite labs situated on board the *Karlissa B* and on each pylon. They carried out all the compliance testing and operated in accordance with documented quality

Fig. 3. The charity walk

procedures following BS 5750. Altogether, 225 000 m³ of structural concrete and 90 000 m³ of mass concrete were placed on the contract, involving the testing of over 48 000 cubes.

Further details of the concrete production on site are given by Mizon et al.[1]

Environmental issues

All teams which successfully prequalified to tender for the project were required to provide proposals for protecting the environment both during and after construction. These were later developed into a detailed statement of action during the parliamentary process and included in the contractual documentation.

Laing-GTM undertook to provide a full-time, independent environmental liaison officer to monitor and control every aspect of the environmental statement. SGS Environment was awarded the contract and prior to the start of construction carried out a series of baseline measurements of rare plants, invertebrates and birds across the site which contained mudflats, salt marshes, ancient reclaimed land, ditches and reens. Large areas, particularly on the Welsh shore, were designated as sites of special scientific interest (SSSIs). These surveys were continued during construction and will continue for a further three years to monitor the effect of the project.

The issues of noise and dust were particularly important in relation to community relations with the local residents. Working hours, noise and dust were all subject to limitation and were monitored on a regular basis to ensure compliance with the limits.

Another area of concern was the quality of fresh water pumped from the British Rail tunnel, emanating from the Great Spring, which runs through rocks under the foundations. This water is used by a number of commercial concerns on the Welsh shore and Whitbread, the major user, established trigger levels for each possible contaminant.

A procedure was put in place to deal with spillages on site to prevent contamination reaching the supply or, failing that, to ensure that contaminated supplies were not used.

Following the completion of construction, a lot of attention is being paid to the reinstatement and replanting of the site construction yards on both shores, which have to be returned to their original use as low-grade agricultural land.

The environmental monitoring will continue for three years after construction to assess the effect of the presence of the operational motorway across the site and to monitor the rate at which the mudflats and salt marshes, disturbed during construction, return to a stable, albeit different, equilibrium condition.

MANAGEMENT OF CONSTRUCTION

Fig. 4. The completed crossing, looking west

Twelve months after completion of the crossing, it is too early to draw any detailed conclusions about the long-term environmental effects but the estuary is slowly establishing a new equilibrium regime and further improvement is expected.

Community relations

The relationship of Laing-GTM with the community, both local and at large, was an important consideration in the management of the construction works. In order to mitigate disruption and inconvenience, a policy of open communication through a series of organizational features was adopted and controlled by a full-time community relations manager.

Local residents, parish and district councillors, sixth-formers from the local schools, environmental agencies and police were invited to be members of the construction consultative groups, one centred on the English side and the other on the Welsh side, within the areas which were most affected.

The groups met quarterly and had the opportunity of full and frank discussions on issues which affected the community they represented and, where necessary, mitigation measures were taken. A progress report was given by the contractor and the future programme explained.

To ensure that communication was effective and that the concerns of the community could be readily expressed, local residents were invited to telephone a named person on a published number to register any complaint. The time, nature and originator of each complaint were logged and the appropriate action taken by the site team. The action was noted and reported back to the originator.

The *Second Severn Crossing News* was edited by the Laing press office and published quarterly. It contained introductions to key project staff, overall statements in regard to construction methods, special features on environmental matters and the influence of the new crossing on the area and a series presenting technical details. Circulation was limited to those areas judged to be the most directly affected by the construction activity. This required in excess of 10 000 copies to be circulated to households at both the English and Welsh ends of the new crossing.

Interest in the project was such that a constant flow of requests for an off-site lecturing service developed with both designer's and contractor's staff spending many days and evenings talking to a variety of interested groups in schools, colleges, village halls, hotels and public houses.

Demand from visitors for 'bridge memorabilia' grew to such an extent that a 'shop' was set up at the riverside public footpath to sell a small selection of keepsakes.

In anticipation of a degree of interest in the project, a visitors' centre was set up to which private visits by interested groups and associations could be made, alternating with days for the general public to view the displays. In the event, interest from private groups was such that weekdays were entirely taken up with pre-arranged visits and additional opening for the general public to visit on summer Sundays was required.

Visitors ranged from primary school students to senior citizens and from laypersons to highly technically qualified engineers from many countries around the world. Over the period of 3½ years some 15 000 people were given a private lecture and site tour and in addition an estimated 60 000 visitors came to view the construction works from the riverside public footpath which ran along the top of the flood defence wall.

Three weeks prior to the opening of the bridge a charity walk (Fig. 3) was organized for 15 000 members of the public to walk across the bridge. This was a unique experience since there is no public footpath along the bridge.

Completion

The bridge was opened on programme by HRH the Prince of Wales on 5 June 1996 and the first public vehicles used the bridge in the early hours of the following morning (Figs 4–6).

Acknowledgements

The authors acknowledge with thanks the assistance received from many of their colleagues in the preparation of this paper, and in particular C. M. Tong of Halcrow-SEEE and P. Iley of Maunsell. Photographs are reproduced with kind permission of Photographic Engineering Services.

Reference

1. MIZON D., KITCHENER N. and KOTRYS L. Concrete for the Second Severn crossing. *Conf. on Concrete in the Service of Mankind, Dundee*, 1996.

Fig. 5 (top). The completed crossing, viewed from the Welsh shore

Fig. 6 (above). The completed crossing, looking north-east, with the first crossing in the background

Second Severn Crossing Group

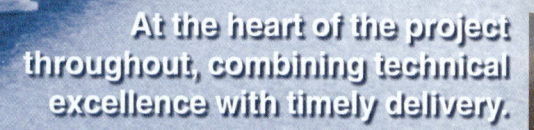

At the heart of the project throughout, combining technical excellence with timely delivery.

1984 to 1987 — Detailed studies into crossing type and location

1987 to 1989 — Preparation of main bridge tender and illustrative design

1989 to 1992 — Assistance in Parliamentary process and detailed design of approach roads

1992 to 1997 — Site supervision and audit

With many thanks to all the team and in special memory of the immense contribution of Brian Richmond

Joint Venture

Maunsell WS Atkins Maunsell WS Atkins

Contact:
Peter Gosling
W S Atkins Consultants Ltd
Woodcote Grove
Ashley Road
EPSOM
Surrey KT18 5BW

Robin Sham
Maunsell Ltd
Maunsell House
160 Croydon Road
BECKENHAM
Kent BR3 4DE

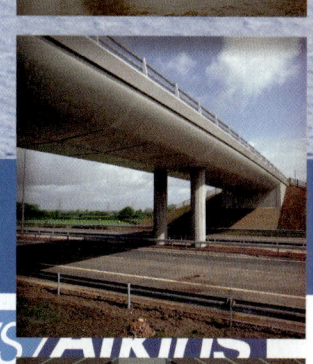

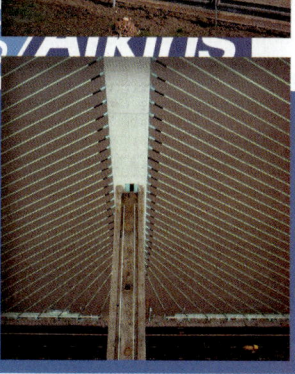

Flint & Neill Partnership
Consulting Civil & Structural Engineers

- More than 25 years involvement with the existing Severn Crossing.
- British Construction Industry Civil Engineering and Supreme Award 1990, for the design and supervision of the strengthening and refurbishment of bridge superstructures.
- Advice on operation, inspection and maintenance procedures.
- Consultants on major bridge schemes worldwide.

The Government's Representative for the operation and maintenance of both Severn crossings.

Stone
Berkeley
Gloucestershire
GL13 9LB

Tel: 01454 260910
Fax: 01454 260784

Also at
21 Dartmouth Street
London SW1H 9BP

And in
Hong Kong

Steelfields Concrete Batching Plant

STEELFIELDS Limited, established in 1956, is a leading authority on the design and manufacture of Concrete Batching and Mixing Plant.

The 'MAJOR' range is based on a modular system which allows tailor-made solutions to satisfy the increasing and diverse demands of the Construction, Readymix and Pre-cast industries.

STEELFIELDS has experience in operating in over 35 countries world-wide and can offer an extensive range of mobile, portable and static Plant together with full ancillary options upon request.

- **Concrete Batching / Mixing Plant**
- **High Speed Pan Mixers**
- **Planetary Mixers**
- **Twin-Shaft Paddle Mixers**
- **Tilting / Horizontal / Split Drum Mixers**
- **Concrete Recycling Systems**
- **Computerised Batching Systems**

Steelfields Limited
Gads Hill, Gillingham, Kent ME7 2RT, United Kingdom
Telephone: 01634 280135 Fax: 01634 280689

STEELFIELDS

The Institution of Civil Engineers is on the Internet now.

Are you?

Now *all* ICE and Thomas Telford Journals are available on-line via our new service **Journals on-line**

http://www.ice.org.uk/journals.html

THE INSTITUTION OF CIVIL ENGINEERS
www.ice.org.uk

Our Service

Contents Update View contents of current issues on-line with this hypertext linked contents database

Abstract Alert Search through all our journal abstracts by title, author or keyword to find papers relevant to your needs

Subscribers can now access **complete issues on-line** that preserve the format of the printed originals and are available **up to 3 weeks** before the printed journal - providing up-to-the-minute information wherever you are

ICEnet also allows you to access a whole range of services including:

The Chamber - An on-line discussion forum **The Library** - With access to the full catalogue on-line **Meetings and Conferences** - Listings of forthcoming events **Professional Development** - Information on Continuing Professional Development **Jobs** - Vacancies from Thomas Telford Recruitment **Students** - A wide range of advice on qualifications, careers, employers and scholarships

Join the electronic revolution NOW!!

For more details about services available via the ICE's website please contact:
The Electronic Publishing Department at
Thomas Telford Publishing, 1 Heron Quay,
London E14 4JD
Telephone: 0171 665 2452
Fax: 0171 538 4101
Email: ttep@ice.org.uk

THE INSTITUTION OF CIVIL ENGINEERS

Thomas Telford